FREE Test Taking Tips DVD Offer

To help us better serve you, we have developed a Test Taking Tips DVD that we would like to give you for FREE. **This DVD covers world-class test taking tips that you can use to be even more successful when you are taking your test.**

All that we ask is that you email us your feedback about your study guide. Please let us know what you thought about it – whether that is good, bad or indifferent.

To get your **FREE Test Taking Tips DVD**, email freedvd@studyguideteam.com with "FREE DVD" in the subject line and the following information in the body of the email:

a. The title of your study guide.

b. Your product rating on a scale of 1-5, with 5 being the highest rating.

c. Your feedback about the study guide. What did you think of it?

d. Your full name and shipping address to send your free DVD.

If you have any questions or concerns, please don't hesitate to contact us at freedvd@studyguideteam.com.

Thanks again!

College Placement Test Study Guide

Test Prep and Practice Exam Questions for the ACCUPLACER, TSI, and Other Placement Exams

[2nd Edition]

TPB Publishing

Written and edited by TPB Publishing.

ISBN 13: 9781628459500
ISBN 10: 1628459506

Table of Contents

Quick Overview

As you draw closer to taking your exam, effective preparation becomes more and more important. Thankfully, you have this study guide to help you get ready. Use this guide to help keep your studying on track and refer to it often.

This study guide contains several key sections that will help you be successful on your exam. The guide contains tips for what you should do the night before and the day of the test. Also included are test-taking tips. Knowing the right information is not always enough. Many well-prepared test takers struggle with exams. These tips will help equip you to accurately read, assess, and answer test questions.

A large part of the guide is devoted to showing you what content to expect on the exam and to helping you better understand that content. In this guide are practice test questions so that you can see how well you have grasped the content. Then, answer explanations are provided so that you can understand why you missed certain questions.

Don't try to cram the night before you take your exam. This is not a wise strategy for a few reasons. First, your retention of the information will be low. Your time would be better used by reviewing information you already know rather than trying to learn a lot of new information. Second, you will likely become stressed as you try to gain a large amount of knowledge in a short amount of time. Third, you will be depriving yourself of sleep. So be sure to go to bed at a reasonable time the night before. Being well-rested helps you focus and remain calm.

Be sure to eat a substantial breakfast the morning of the exam. If you are taking the exam in the afternoon, be sure to have a good lunch as well. Being hungry is distracting and can make it difficult to focus. You have hopefully spent lots of time preparing for the exam. Don't let an empty stomach get in the way of success!

When travelling to the testing center, leave earlier than needed. That way, you have a buffer in case you experience any delays. This will help you remain calm and will keep you from missing your appointment time at the testing center.

Be sure to pace yourself during the exam. Don't try to rush through the exam. There is no need to risk performing poorly on the exam just so you can leave the testing center early. Allow yourself to use all of the allotted time if needed.

Remain positive while taking the exam even if you feel like you are performing poorly. Thinking about the content you should have mastered will not help you perform better on the exam.

Once the exam is complete, take some time to relax. Even if you feel that you need to take the exam again, you will be well served by some down time before you begin studying again. It's often easier to convince yourself to study if you know that it will come with a reward!

Test-Taking Strategies

1. Predicting the Answer

When you feel confident in your preparation for a multiple-choice test, try predicting the answer before reading the answer choices. This is especially useful on questions that test objective factual knowledge. By predicting the answer before reading the available choices, you eliminate the possibility that you will be distracted or led astray by an incorrect answer choice. You will feel more confident in your selection if you read the question, predict the answer, and then find your prediction among the answer choices. After using this strategy, be sure to still read all of the answer choices carefully and completely. If you feel unprepared, you should not attempt to predict the answers. This would be a waste of time and an opportunity for your mind to wander in the wrong direction.

2. Reading the Whole Question

Too often, test takers scan a multiple-choice question, recognize a few familiar words, and immediately jump to the answer choices. Test authors are aware of this common impatience, and they will sometimes prey upon it. For instance, a test author might subtly turn the question into a negative, or he or she might redirect the focus of the question right at the end. The only way to avoid falling into these traps is to read the entirety of the question carefully before reading the answer choices.

3. Looking for Wrong Answers

Long and complicated multiple-choice questions can be intimidating. One way to simplify a difficult multiple-choice question is to eliminate all of the answer choices that are clearly wrong. In most sets of answers, there will be at least one selection that can be dismissed right away. If the test is administered on paper, the test taker could draw a line through it to indicate that it may be ignored; otherwise, the test taker will have to perform this operation mentally or on scratch paper. In either case, once the obviously incorrect answers have been eliminated, the remaining choices may be considered. Sometimes identifying the clearly wrong answers will give the test taker some information about the correct answer. For instance, if one of the remaining answer choices is a direct opposite of one of the eliminated answer choices, it may well be the correct answer. The opposite of obviously wrong is obviously right! Of course, this is not always the case. Some answers are obviously incorrect simply because they are irrelevant to the question being asked. Still, identifying and eliminating some incorrect answer choices is a good way to simplify a multiple-choice question.

4. Don't Overanalyze

Anxious test takers often overanalyze questions. When you are nervous, your brain will often run wild, causing you to make associations and discover clues that don't actually exist. If you feel that this may be a problem for you, do whatever you can to slow down during the test. Try taking a deep breath or counting to ten. As you read and consider the question, restrict yourself to the particular words used by the author. Avoid thought tangents about what the author *really* meant, or what he or she was *trying* to say. The only things that matter on a multiple-choice test are the words that are actually in the question. You must avoid reading too much into a multiple-choice question, or supposing that the writer meant something other than what he or she wrote.

5. No Need for Panic

It is wise to learn as many strategies as possible before taking a multiple-choice test, but it is likely that you will come across a few questions for which you simply don't know the answer. In this situation, avoid panicking. Because most multiple-choice tests include dozens of questions, the relative value of a single wrong answer is small. As much as possible, you should compartmentalize each question on a multiple-choice test. In other words, you should not allow your feelings about one question to affect your success on the others. When you find a question that you either don't understand or don't know how to answer, just take a deep breath and do your best. Read the entire question slowly and carefully. Try rephrasing the question a couple of different ways. Then, read all of the answer choices carefully. After eliminating obviously wrong answers, make a selection and move on to the next question.

6. Confusing Answer Choices

When working on a difficult multiple-choice question, there may be a tendency to focus on the answer choices that are the easiest to understand. Many people, whether consciously or not, gravitate to the answer choices that require the least concentration, knowledge, and memory. This is a mistake. When you come across an answer choice that is confusing, you should give it extra attention. A question might be confusing because you do not know the subject matter to which it refers. If this is the case, don't eliminate the answer before you have affirmatively settled on another. When you come across an answer choice of this type, set it aside as you look at the remaining choices. If you can confidently assert that one of the other choices is correct, you can leave the confusing answer aside. Otherwise, you will need to take a moment to try to better understand the confusing answer choice. Rephrasing is one way to tease out the sense of a confusing answer choice.

7. Your First Instinct

Many people struggle with multiple-choice tests because they overthink the questions. If you have studied sufficiently for the test, you should be prepared to trust your first instinct once you have carefully and completely read the question and all of the answer choices. There is a great deal of research suggesting that the mind can come to the correct conclusion very quickly once it has obtained all of the relevant information. At times, it may seem to you as if your intuition is working faster even than your reasoning mind. This may in fact be true. The knowledge you obtain while studying may be retrieved from your subconscious before you have a chance to work out the associations that support it. Verify your instinct by working out the reasons that it should be trusted.

8. Key Words

Many test takers struggle with multiple-choice questions because they have poor reading comprehension skills. Quickly reading and understanding a multiple-choice question requires a mixture of skill and experience. To help with this, try jotting down a few key words and phrases on a piece of scrap paper. Doing this concentrates the process of reading and forces the mind to weigh the relative importance of the question's parts. In selecting words and phrases to write down, the test taker thinks about the question more deeply and carefully. This is especially true for multiple-choice questions that are preceded by a long prompt.

9. Subtle Negatives

One of the oldest tricks in the multiple-choice test writer's book is to subtly reverse the meaning of a question with a word like *not* or *except*. If you are not paying attention to each word in the question, you can easily be led astray by this trick. For instance, a common question format is, "Which of the following is...?" Obviously, if the question instead is, "Which of the following is not...?," then the answer will be quite different. Even worse, the test makers are aware of the potential for this mistake and will include one answer choice that would be correct if the question were not negated or reversed. A test taker who misses the reversal will find what he or she believes to be a correct answer and will be so confident that he or she will fail to reread the question and discover the original error. The only way to avoid this is to practice a wide variety of multiple-choice questions and to pay close attention to each and every word.

10. Reading Every Answer Choice

It may seem obvious, but you should always read every one of the answer choices! Too many test takers fall into the habit of scanning the question and assuming that they understand the question because they recognize a few key words. From there, they pick the first answer choice that answers the question they believe they have read. Test takers who read all of the answer choices might discover that one of the latter answer choices is actually *more* correct. Moreover, reading all of the answer choices can remind you of facts related to the question that can help you arrive at the correct answer. Sometimes, a misstatement or incorrect detail in one of the latter answer choices will trigger your memory of the subject and will enable you to find the right answer. Failing to read all of the answer choices is like not reading all of the items on a restaurant menu: you might miss out on the perfect choice.

11. Spot the Hedges

One of the keys to success on multiple-choice tests is paying close attention to every word. This is never truer than with words like almost, most, some, and sometimes. These words are called "hedges" because they indicate that a statement is not totally true or not true in every place and time. An absolute statement will contain no hedges, but in many subjects, the answers are not always straightforward or absolute. There are always exceptions to the rules in these subjects. For this reason, you should favor those multiple-choice questions that contain hedging language. The presence of qualifying words indicates that the author is taking special care with his or her words, which is certainly important when composing the right answer. After all, there are many ways to be wrong, but there is only one way to be right! For this reason, it is wise to avoid answers that are absolute when taking a multiple-choice test. An absolute answer is one that says things are either all one way or all another. They often include words like *every*, *always*, *best*, and *never*. If you are taking a multiple-choice test in a subject that doesn't lend itself to absolute answers, be on your guard if you see any of these words.

12. Long Answers

In many subject areas, the answers are not simple. As already mentioned, the right answer often requires hedges. Another common feature of the answers to a complex or subjective question are qualifying clauses, which are groups of words that subtly modify the meaning of the sentence. If the question or answer choice describes a rule to which there are exceptions or the subject matter is complicated, ambiguous, or confusing, the correct answer will require many words in order to be expressed clearly and accurately. In essence, you should not be deterred by answer choices that seem excessively long. Oftentimes, the author of the text will not be able to write the correct answer without offering some qualifications and modifications. Your job is to read the answer choices thoroughly and completely and to select the one that most accurately and precisely answers the question.

13. Restating to Understand

Sometimes, a question on a multiple-choice test is difficult not because of what it asks but because of how it is written. If this is the case, restate the question or answer choice in different words. This process serves a couple of important purposes. First, it forces you to concentrate on the core of the question. In order to rephrase the question accurately, you have to understand it well. Rephrasing the question will concentrate your mind on the key words and ideas. Second, it will present the information to your mind in a fresh way. This process may trigger your memory and render some useful scrap of information picked up while studying.

14. True Statements

Sometimes an answer choice will be true in itself, but it does not answer the question. This is one of the main reasons why it is essential to read the question carefully and completely before proceeding to the answer choices. Too often, test takers skip ahead to the answer choices and look for true statements. Having found one of these, they are content to select it without reference to the question above. Obviously, this provides an easy way for test makers to play tricks. The savvy test taker will always read the entire question before turning to the answer choices. Then, having settled on a correct answer choice, he or she will refer to the original question and ensure that the selected answer is relevant. The mistake of choosing a correct-but-irrelevant answer choice is especially common on questions related to specific pieces of objective knowledge. A prepared test taker will have a wealth of factual knowledge at his or her disposal, and should not be careless in its application.

15. No Patterns

One of the more dangerous ideas that circulates about multiple-choice tests is that the correct answers tend to fall into patterns. These erroneous ideas range from a belief that B and C are the most common right answers, to the idea that an unprepared test-taker should answer "A-B-A-C-A-D-A-B-A." It cannot be emphasized enough that pattern-seeking of this type is exactly the WRONG way to approach a multiple-choice test. To begin with, it is highly unlikely that the test maker will plot the correct answers according to some predetermined pattern. The questions are scrambled and delivered in a random order. Furthermore, even if the test maker was following a pattern in the assignation of correct answers, there is no reason why the test taker would know which pattern he or she was using. Any attempt to discern a pattern in the answer choices is a waste of time and a distraction from the real work of taking the test. A test taker would be much better served by extra preparation before the test than by reliance on a pattern in the answers.

FREE DVD OFFER

Don't forget that doing well on your exam includes both understanding the test content and understanding how to use what you know to do well on the test. We offer a completely FREE Test Taking Tips DVD that covers world class test taking tips that you can use to be even more successful when you are taking your test.

All that we ask is that you email us your feedback about your study guide. To get your **FREE Test Taking Tips DVD**, email freedvd@studyguideteam.com with "FREE DVD" in the subject line and the following information in the body of the email:

- The title of your study guide.
- Your product rating on a scale of 1-5, with 5 being the highest rating.
- Your feedback about the study guide. What did you think of it?
- Your full name and shipping address to send your free DVD.

Introduction to College Placement Exams

Function of the Test

College placement exams are used to measure a student's readiness for college-level coursework. They are used by some colleges and high schools to determine placement of students in programs appropriate to the students' skill level. Schools can also use a college placement exam to identify specific areas in which the students need improvement. Scores are generally used only by the college or high school the student is already attending for placement and instruction at that school.

Some students can be exempt from these tests if they met the minimum score on the SAT, ACT or statewide high school test or if they have already completed certain college-level courses.

Test Administration

College placement exams will often be administered at testing centers located at universities. Sometimes they will be administered at a student's high school. Retesting is sometimes acceptable and some colleges and universities also host workshops to better prepare a student to retake the exam. It is advised to contact the university or college where a candidate wishes to take the test to receive more information on retesting and workshops. Advisors and counselors at a particular institution can better help a candidate with any further questions they may have.

Test Format

Test formats vary depending on the test but they often focus on:

- Mathematics
- Reading
- Writing

This study guide covers content for those three areas and explains the concepts that are often tested on these exams.

Scoring

The minimum passing score will vary by test and/or institution. Be sure to check with your college or future college to find out about their exact requirements

Mathematics

Arithmetic

Addition with Whole Numbers and Fractions

Addition combines two quantities together. With whole numbers, this is taking two sets of things and merging them into one, then counting the result. For example, 4 + 3 = 7. When adding numbers, the order does not matter: 3 + 4 = 7, also. Longer lists of whole numbers can also be added together. The result of adding numbers is called the *sum*.

With fractions, the number on top is the *numerator*, and the number on the bottom is the *denominator*. To add fractions, the denominator must be the same—a *common denominator*. To find a common denominator, the existing numbers on the bottom must be considered, and the lowest number they will both multiply into must be determined. Consider the following equation:

$$\frac{1}{3}+\frac{5}{6}=?$$

The numbers 3 and 6 both multiply into 6. Three can be multiplied by 2, and 6 can be multiplied by 1. The top and bottom of each fraction must be multiplied by the same number. Then, the numerators are added together to get a new numerator. The following equation is the result:

$$\frac{1}{3}+\frac{5}{6}=\frac{2}{6}+\frac{5}{6}=\frac{7}{6}$$

Subtraction with Whole Numbers and Fractions

Subtraction is taking one quantity away from another, so it is the opposite of addition. The expression 4 – 3 means taking 3 away from 4. So, 4 – 3 = 1. In this case, the order matters, since it entails taking one quantity away from the other, rather than just putting two quantities together. The result of subtraction is also called the *difference*.

To subtract fractions, the denominator must be the same. Then, subtract the numerators together to get a new numerator. Here is an example:

$$\frac{1}{3}-\frac{5}{6}=\frac{2}{6}-\frac{5}{6}=\frac{-3}{6}=-\frac{1}{2}$$

Multiplication with Whole Numbers and Fractions

Multiplication is a kind of repeated addition. The expression 4×5 is taking four sets, each of them having five things in them, and putting them all together. That means $4 \times 5 = 5 + 5 + 5 + 5 = 20$. As with addition, the order of the numbers does not matter. The result of a multiplication problem is called the *product*.

To multiply fractions, the numerators are multiplied to get the new numerator, and the denominators are multiplied to get the new denominator:

$$\frac{1}{3} \times \frac{5}{6} = \frac{1 \times 5}{3 \times 6} = \frac{5}{18}$$

When multiplying fractions, common factors can *cancel* or *divide into one another*, when factors that appear in the numerator of one fraction and the denominator of the other fraction. Here is an example:

$$\frac{1}{3} \times \frac{9}{8} = \frac{1}{1} \times \frac{3}{8} = 1 \times \frac{3}{8} = \frac{3}{8}$$

The numbers 3 and 9 have a common factor of 3, so that factor can be divided out.

Division with Whole Numbers and Fractions

Division is the opposite of multiplication. With whole numbers, it means splitting up one number into sets of equal size. For example, $16 \div 8$ is the number of sets of eight things that can be made out of sixteen things. Thus, $16 \div 8 = 2$. As with subtraction, the order of the numbers will make a difference, here. The answer to a division problem is called the *quotient*, while the number in front of the division sign is called the *dividend*, and the number behind the division sign is called the *divisor*.

To divide fractions, the first fraction must be multiplied with the reciprocal of the second fraction. The *reciprocal* of the fraction $\frac{x}{y}$ is the fraction $\frac{y}{x}$. Here is an example:

$$\frac{1}{3} \div \frac{5}{6} = \frac{1}{3} \times \frac{6}{5} = \frac{6}{15} = \frac{2}{5}$$

Recognizing Equivalent Fractions and Mixed Numbers

The value of a fraction does not change if multiplying or dividing both the numerator and the denominator by the same number (other than 0). In other words, $\frac{x}{y} = \frac{a \times x}{a \times y} = \frac{x \div a}{y \div a}$, as long as *a* is not 0. This means that $\frac{2}{5} = \frac{4}{10}$, for example. If *x* and *y* are integers that have no common factors, then the fraction is said to be *simplified*. This means $\frac{2}{5}$ is simplified, but $\frac{4}{10}$ is not.

Often when working with fractions, the fractions need to be rewritten so that they all share a single denominator—this is called finding a *common denominator* for the fractions. Using two fractions, $\frac{a}{b}$ and $\frac{c}{d}$, the numerator and denominator of the left fraction can be multiplied by *d*, while the numerator and denominator of the right fraction can be multiplied by *b*. This provides the fractions $\frac{a \times d}{b \times d}$ and $\frac{c \times b}{d \times b}$ with the common denominator $b \times d$.

A fraction whose numerator is smaller than its denominator is called a *proper fraction*. A fraction whose numerator is bigger than its denominator is called an *improper fraction*. These numbers can be rewritten as a combination of integers and fractions, called a *mixed number*. For example, $\frac{6}{5} = \frac{5}{5} + \frac{1}{5} = 1 + \frac{1}{5}$, and can be written as $1\frac{1}{5}$.

Estimating

Estimation is finding a value that is close to a solution but is not the exact answer. For example, if there are values in the thousands to be multiplied, then each value can be estimated to the nearest thousand

and the calculation performed. This value provides an approximate solution that can be determined very quickly.

Recognition of Decimals

The *decimal system* is a way of writing out numbers that uses ten different numerals: 0, 1, 2, 3, 4, 5, 6, 7, 8, and 9. This is also called a “base ten” or “base 10” system. Other bases are also used. For example, computers work with a base of 2. This means they only use the numerals 0 and 1.

The *decimal place* denotes how far to the right of the decimal point a numeral is. The first digit to the right of the decimal point is in the *tenths* place. The next is the *hundredths*. The third is the *thousandths*.

So, 3.142 has a 1 in the tenths place, a 4 in the hundredths place, and a 2 in the thousandths place.

The *decimal point* is a period used to separate the *ones* place from the *tenths* place when writing out a number as a decimal.

A *decimal number* is a number written out with a decimal point instead of as a fraction, for example, 1.25 instead of $\frac{5}{4}$. Depending on the situation, it can sometimes be easier to work with fractions and sometimes easier to work with decimal numbers.

A decimal number is *terminating* if it stops at some point. It is called *repeating* if it never stops but repeats a pattern over and over. It is important to note that every rational number can be written as a terminating decimal or as a repeating decimal.

Addition with Decimals

To add decimal numbers, each number in columns needs to be lined up by the decimal point. For each number being added, the zeros to the right of the last number need to be filled in so that each of the numbers has the same number of places to the right of the decimal. Then, the columns can be added together. Here is an example of 2.45 + 1.3 + 8.891 written in column form:

$$\begin{array}{r} 2.450 \\ 1.300 \\ \underline{+\ 8.891} \end{array}$$

Zeros have been added in the columns so that each number has the same number of places to the right of the decimal.

Added together, the correct answer is 12.641:

$$\begin{array}{r} 2.450 \\ 1.300 \\ \underline{+\ 8.891} \\ 12.641 \end{array}$$

Subtraction with Decimals

Subtracting decimal numbers is the same process as adding decimals. Here is 7.89 – 4.235 written in column form:

$$\begin{array}{r} 7.890 \\ \underline{-\ 4.235} \\ 3.655 \end{array}$$

A zero has been added in the column so that each number has the same number of places to the right of the decimal.

Multiplication with Decimals

The simplest way to multiply decimals is to calculate the product as if the decimals are not there, then count the number of decimal places in the original problem. Use that total to place the decimal the same number of places over in your answer, counting from right to left. For example, 0.5 x 1.25 can be rewritten and multiplied as 5 x 125, which equals 625. Then the decimal is added three places from the right for .625.

The final answer will have the same number of decimal *points* as the total number of decimal *places* in the problem. The first number has one decimal place, and the second number has two decimal places. Therefore, the final answer will contain three decimal places:

0.5 x 1.25 = 0.625

Division with Decimals

Dividing a decimal by a whole number entails using long division first by ignoring the decimal point. Then, the decimal point is moved the number of places given in the problem.

For example, $6.8 \div 4$ can be rewritten as $68 \div 4$, which is 17. There is one non-zero integer to the right of the decimal point, so the final solution would have one decimal place to the right of the solution. In this case, the solution is 1.7.

Dividing a decimal by another decimal requires changing the divisor to a whole number by moving its decimal point. The decimal place of the dividend should be moved by the same number of places as the divisor. Then, the problem is the same as dividing a decimal by a whole number.

For example, $5.72 \div 1.1$ has a divisor with one decimal point in the denominator. The expression can be rewritten as $57.2 \div 11$ by moving each number one decimal place to the right to eliminate the decimal. The long division can be completed as $572 \div 11$ with a result of 52. Since there is one non-zero integer to the right of the decimal point in the problem, the final solution is 5.2.

In another example, $8 \div 0.16$ has a divisor with two decimal points in the denominator. The expression can be rewritten as $800 \div 16$ by moving each number two decimal places to the right to eliminate the decimal in the divisor. The long division can be completed with a result of 50.

Fraction and Percent Equivalencies

The word *percent* comes from the Latin phrase for "per one hundred." A *percent* is a way of writing out a fraction. It is a fraction with a denominator of 100. Thus, $65\% = \frac{65}{100}$.

To convert a fraction to a percent, the denominator is written as 100. For example, $\frac{3}{5} = \frac{60}{100} = 60\%$.

In converting a percent to a fraction, the percent is written with a denominator of 100, and the result is simplified. For example, $30\% = \frac{30}{100} = \frac{3}{10}$.

Percent Problems

The basic percent equation is the following:

$$\frac{is}{of} = \frac{\%}{100}$$

The placement of numbers in the equation depends on what the question asks.

Example 1

Find 40% of 80.

Basically, the problem is asking, "What is 40% of 80?" The 40% is the percent, and 80 is the number to find the percent "of." The equation is:

$$\frac{x}{80} = \frac{40}{100}$$

Solving the equation by cross-multiplication, the problem becomes 100x = 80(40). Solving for x gives the answer: x = 32.

Example 2

What percent of 100 is 20?

The 20 fills in the "is" portion, while 100 fills in the "of." The question asks for the percent, so that will be x, the unknown. The following equation is set up:

$$\frac{20}{100} = \frac{x}{100}$$

Cross-multiplying yields the equation 100x = 20(100). Solving for x gives the answer of 20%.

Example 3

30% of what number is 30?

The following equation uses the clues and numbers in the problem:

$$\frac{30}{x} = \frac{30}{100}$$

Cross-multiplying results in the equation 30(100) = 30x. Solving for x gives the answer x = 100.

Problems Involving Estimation

Sometimes when multiplying numbers, the result can be estimated by *rounding*. For example, to estimate the value of 11.2×2.01, each number can be rounded to the nearest integer. This will yield a result of 22.

Rate, Percent, and Measurement Problems

A *ratio* compares the size of one group to the size of another. For example, there may be a room with 4 tables and 24 chairs. The ratio of tables to chairs is $4: 24$. Such ratios behave like fractions in that both sides of the ratio by the same number can be multiplied or divided. Thus, the ratio 4:24 is the same as the ratio 2:12 and 1:6.

One quantity is *proportional* to another quantity if the first quantity is always some multiple of the second. For instance, the distance travelled in five hours is always five times to the speed as travelled. The distance is proportional to speed in this case.

One quantity is *inversely proportional* to another quantity if the first quantity is equal to some number divided by the second quantity. The time it takes to travel one hundred miles will be given by 100 divided by the speed travelled. The time is inversely proportional to the speed.

When dealing with word problems, there is no fixed series of steps to follow, but there are some general guidelines to use. It is important that the quantity to be found is identified. Then, it can be determined how the given values can be used and manipulated to find the final answer.

Example 1

Jana wants to travel to visit Alice, who lives one hundred and fifty miles away. If she can drive at fifty miles per hour, how long will her trip take?

The quantity to find is the *time* of the trip. The time of a trip is given by the distance to travel divided by the speed to be traveled. The problem determines that the distance is one hundred and fifty miles, while the speed is fifty miles per hour. Thus, 150 divided by 50 is $150 \div 50 = 3$. Because *miles* and *miles per hour* are the units being divided, the miles cancel out. The result is 3 hours.

Example 2

Bernard wishes to paint a wall that measures twenty feet wide by eight feet high. It costs ten cents to paint one square foot. How much money will Bernard need for paint?

The final quantity to compute is the *cost* to paint the wall. This will be ten cents ($0.10) for each square foot of area needed to paint. The area to be painted is unknown, but the dimensions of the wall are given; thus, it can be calculated.

The dimensions of the wall are 20 feet wide and 8 feet high. Since the area of a rectangle is length multiplied by width, the area of the wall is 8 x 20 = 160 square feet. Multiplying 0.1 x 160 yields $16 as the cost of the paint.

The *average* or *mean* of a collection of numbers is given by adding those numbers together and then dividing by the total number of values. A *weighted average* or *weighted mean* is given by adding the numbers multiplied by their weights, then dividing by the sum of the weights:

$$\frac{w_1x_1 + w_2x_2 + w_3x_3 \dots + w_nx_n}{w_1 + w_2 + w_3 + \cdots + w_n}$$

An *ordinary average* is a weighted average where all the weights are 1.

Fractions and Word Problems

Work word problems are examples of people working together in a situation that uses fractions.

Example

One painter can paint a designated room in 6 hours, and a second painter can paint the same room in 5 hours. How long will it take them to paint the room if they work together?

The first painter paints $\frac{1}{6}$of the room in an hour, and the second painter paints $\frac{1}{5}$of the room in an hour.

Together, they can paint $\frac{1}{x}$ of the room in an hour. The equation is the sum of the painters rate equal to the total job or $\frac{1}{6} + \frac{1}{5} = \frac{1}{x}$.

The equation can be solved by multiplying all terms by a common denominator of 30*x* with a result of $5x + 6x = 30$.

The left side can be added together to get 11*x*, and then divide by 11 for a solution of $\frac{30}{11}$or about 2.73 hours.

Distribution of a Quantity into its Fractional Parts

A quantity may be broken into its fractional parts. For example, a toy box holds three types of toys for kids. $\frac{1}{3}$ of the toys are Type A and $\frac{1}{4}$ of the toys are Type B. With that information, how many Type C toys are there?

First, the sum of Type A and Type B must be determined by finding a common denominator to add the fractions. The lowest common multiple is 12, so that is what will be used. The sum is $\frac{1}{3} + \frac{1}{4} = \frac{4}{12} + \frac{3}{12} = \frac{7}{12}$.

This value is subtracted from 1 to find the number of Type C toys. The value is subtracted from 1 because 1 represents a whole. The calculation is $1 - \frac{7}{12} = \frac{12}{12} - \frac{7}{12} = \frac{5}{12}$. This means that $\frac{5}{12}$ of the toys are Type C. To check the answer, add all fractions together, and the result should be 1.

Elementary Algebra and Functions

Computation with Integers and Negative Rational Numbers

Integers are the whole numbers together with their negatives. They include numbers like 5, 24, 0, -6, and 15. They not include fractions or numbers that have digits after the decimal point.

Rational numbers are all numbers that can be written as a fraction using integers. A *fraction* is written as $\frac{x}{y}$ and represents the quotient of *x* being divided by *y*. More practically, it means dividing the whole into *y* equal parts, then taking *x* of those parts.

Examples of rational numbers include $\frac{1}{2}$ and $\frac{5}{4}$. The number on the top is called the *numerator*, and the number on the bottom is called the *denominator*. Because every integer can be written as a fraction with a denominator of 1, (e.g. $\frac{3}{1} = 3$), every integer is also a rational number.

When adding integers and negative rational numbers, there are some basic rules to determine if the solution is negative or positive:

Adding two positive numbers results in a positive number: 3.3 + 4.8 = 8.1.

Adding two negative numbers results in a negative number: (-8) + (-6) = -14.

Adding one positive and one negative number requires taking the absolute values and finding the difference between them. Then, the sign of the number that has the higher absolute value for the final solution is used.

For example, (-9) + 11, has a difference of absolute values of 2. The final solution is 2 because 11 has the higher absolute value. Another example is 9 + (-11), which has a difference of absolute values of 2. The final solution is -2 because 11 has the higher absolute value.

When subtracting integers and negative rational numbers, one has to change the problem to adding the opposite and then apply the rules of addition.

Subtracting two positive numbers is the same as adding one positive and one negative number.

For example, 4.9 – 7.1 is the same as 4.9 + (-7.1). The solution is -2.2 since the absolute value of -7.1 is greater. Another example is 8.5 – 6.4 which is the same as 8.5 + (-6.4). The solution is 2.1 since the absolute value of 8.5 is greater.

Subtracting a positive number from a negative number results in negative value.

For example, (-12) – 7 is the same as (-12) + (-7) with a solution of -19.

Subtracting a negative number from a positive number results in a positive value.

For example, 12 – (-7) is the same as 12 + 7 with a solution of 19.

For multiplication and division of integers and rational numbers, if both numbers are positive or both numbers are negative, the result is a positive value.

For example, (-1.7)(-4) has a solution of 6.8 since both numbers are negative values.

If one number is positive and another number is negative, the result is a negative value.

For example, (-15)/5 has a solution of -3 since there is one negative number.

The Use of Absolute Values

The *absolute value* represents the distance a number is from 0. The *absolute value symbol* is | | with a number between the bars. The |10| = 10 and the |-10| = 10.

When simplifying an algebraic expression, the value of the absolute value expression is determined first, much like parenthesis in the order of operations. See the example below:

$$|8 - 12| + 5 = |-4| + 5 = 4 + 5 = 9$$

Exponents and Roots

An *exponent* is written as a^b. In this expression, *a* is called the *base* and *b* is called the *exponent*. It is properly stated that *a* is raised to the *n*-th power. Therefore, in the expression 2^3, the exponent is 3, while the base is 2. Such an expression is called an *exponential expression*. Note that when the exponent is 2, it is called *squaring* the base, and when it is 3, it is called *cubing* the base.

When the exponent is a positive integer, this indicates the base is multiplied by itself the number of times written in the exponent. So, in the expression 2^3, multiply 2 by itself with 3 copies of 2: $2^3 = 2 \times 2 \times 2 = 8$. One thing to notice is that, for positive integers *n* and *m*, $a^n a^m = a^{n+m}$ is a rule. In order to make this rule be true for an integer of 0, $a^0 = 1$, so that $a^n a^0 = a^{n+0} = a^n$. And, in order to make this rule be true for negative exponents, $a^{-n} = \frac{1}{a^n}$.

Another rule for simplifying expressions with exponents is shown by the following equation: $(a^m)^n = a^{mn}$. This is true for fractional exponents as well. So, for a positive integer, define $a^{\frac{1}{n}}$ to be the number that, when raised to the *n*-th power, provides *a*. In other words, $(a^{\frac{1}{n}})^n = a$ is the desired equation. It should be noted that $a^{\frac{1}{n}}$ is the *n*-th root of *a*. This also can be written as $a^{\frac{1}{n}} = \sqrt[n]{a}$. The symbol on the right-hand side of this equation is called a *radical*. If the root is left out, assume that the 2nd root should be taken, also called the *square* root: $a^{\frac{1}{2}} = \sqrt[2]{a} = \sqrt{a}$. Additionally, $\sqrt[3]{a}$ is also called the *cube* root.

Note that when multiple roots exist, $a^{\frac{1}{n}}$ is defined to be the *positive* root. So, $4^{\frac{1}{2}} = 2$. Also note that negative numbers do not have even roots in the real numbers.

This also enables finding exponents for any rational number: $a^{\frac{m}{n}} = (a^{\frac{1}{n}})^m = (a^m)^{\frac{1}{n}}$. In fact, the exponent can be any real number. In general, the following rules for exponents should be used for any numbers $a, b, m,$ and n.

- $a^1 = a$.
- $1^a = 1$.
- $a^0 = 1$.
- $a^m a^n = a^{m+n}$.

- $\frac{a^m}{a^n} = a^{m-n}$
- $(a^m)^n = a^{m \times n}$.
- $(ab)^m = a^m b^m$.
- $(\frac{a}{b})^m = \frac{a^m}{b^m}$.

As an example of applying these rules, consider the problem of simplifying the expression $(3x^2y)^3(2xy^4)$. Start by simplifying the left term using the sixth rule listed. Applying this rule yields the following expression: $27x^6y^3(2xy^4)$. The exponents can now be combined with base *x* and the exponents with base *y*. Multiply the coefficients to yield $54x^7y^7$.

Solving Equations with Exponents and Roots

Here are some of the most important properties of exponents and roots: if *n* is an integer, and if $a^n = b^n$, then $a = b$ if *n* is odd; but $a = \pm b$ if *n* is even. Similarly, if the roots of two things are equal, $\sqrt[n]{a} = \sqrt[n]{b}$, then $a = b$. This means that when starting with a true equation, both sides of that equation can be raised to a given power to obtain another true equation. Beware that when an even-powered root is taken on both sides of the equation, a $\pm$ in the result. For example, given the equation $x^2 = 16$, take the square root of both sides to solve for x. This results in the answer $x = \pm 4$ because $(-4)^2 = 16$ and $(4)^2 = 16$.

Another property is that if $a^n = a^m$, then $n = m$. This is true for any real numbers *n* and *m*.

For solving the equation $\sqrt{x+2} - 1 = 3$, start by moving the -1 over to the right-hand side. This is performed by adding 1 to both sides, which yields $\sqrt{x+2} = 4$. Now, square both sides, but remember that by squaring both sides, the signs are irrelevant. This yields $x + 2 = 16$, which simplifies to give $x = 14$.

Now consider the problem $(x+1)^4 = 16$. To solve this, take the 4th root of both sides, which means an ambiguity in the sign will be introduced because it is an even root: $\sqrt[4]{(x+1)^4} = \pm\sqrt[4]{16}$. The right-hand side is 2, since $2^4 = 16$. Therefore, $x + 1 = \pm 2$ or $x = -1 \pm 2$. Thus, the two possible solutions are $x = -3$ and $x = 1$.

Remember that when solving equations, the answer can be checked by plugging the solution back into the problem to make a true statement.

In sum, there are a few rules for working with exponents. For any numbers a, b, m, n, the following hold true:

$$a^1 = a$$

$$1^a = 1$$

$$a^0 = 1$$

$$a^m \times a^n = a^{m+n}$$

$$a^m \div a^n = a^{m-n}$$

$$(a^m)^n = a^{m \times n}$$

$$(a \times b)^m = a^m \times b^m$$

$$(a \div b)^m = a^m \div b^m$$

Any number, including a fraction, can be an exponent. The same rules apply.

Order of Operations

When working with complicated expressions, parentheses are used to indicate in which order to perform operations. However, to avoid having too many parentheses in an expression, here are some basic rules concerning the proper order to perform operations when not otherwise specified.

1. Parentheses: always perform operations inside parentheses first, regardless of what those operations are
2. Exponents
3. Multiplication and Division
4. Addition and Subtraction

For #3 & #4, work these from left to right. So, if there a subtraction problem and then an addition problem, the subtraction problem will be worked first.

Note that multiplication and division are performed from left to right as they appear in the expression or equation. Addition and subtraction also are performed from left to right as they appear.

To help remember this, many students like to use the mnemonic PEMDAS. Some students associate this word with a phrase to help them, such as "Pirates Eat Many Donuts at Sea." Here is a quick example:

Evaluate $2^2 \times (3 - 1) \div 2 + 3$.

Parenthesis: $2^2 \times 2 \div 2 + 3$.

Exponents: $4 \times 2 \div 2 + 3$

Multiply: $8 \div 2 + 3$.

Divide: $4 + 3$.

Addition: 7

Relations and Functions

First, it's important to understand the definition of a *relation*. Given two variables, *x* and *y*, which stand for unknown numbers, a *relation* between *x* and *y* is an object that splits all of the pairs (*x*, *y*) into those for which the relation is true and those for which it is false. For example, consider the relation of $x^2 = y^2$. This relationship is true for the pair (1, 1) and for the pair (-2, 2), but false for (2, 3). Another example of a relation is $x \leq y$. This is true whenever *x* is less than or equal to *y*.

A *function* is a special kind of relation where, for each value of *x*, there is only a single value of *y* that satisfies the relation. So, $x^2 = y^2$ is *not* a function because in this case, if *x* is 1, *y* can be either 1 or -1: the pair (1, 1) and (1, -1) both satisfy the relation. More generally, for this relation, any pair of the form $(a, \pm a)$ will satisfy it. On the other hand, consider the following relation: $y = x^2 + 1$. This is a function because for each value of *x*, there is a unique value of *y* that satisfies the relation. Notice, however, there are multiple values of *x* that give us the same value of *y*. This is perfectly acceptable for a function. Therefore, *y* is a function of *x*.

To determine if a relation is a function, check to see if every *x* value has a unique corresponding *y* value.

A function can be viewed as an object that has *x* as its input and outputs a unique *y*-value. It is sometimes convenient to express this using *function notation*, where the function itself is given a name, often *f*. To emphasize that *f* takes *x* as its input, the function is written as $f(x)$. In the above example, the equation could be rewritten as $f(x) = x^2 + 1$. To write the value that a function yields for some specific value of *x*, that value is put in place of *x* in the function notation. For example, $f(3)$ means the value that the function outputs when the input value is 3. If $f(x) = x^2 + 1$, then $f(3) = 3^2 + 1 = 10$.

A function can also be viewed as a table of pairs (*x*, *y*), which lists the value for *y* for each possible value of *x*.

The set of all possible values for *x* in $f(x)$ is called the *domain* of the function, and the set of all possible outputs is called the *range* of the function. Note that usually the domain is assumed to be all real numbers, except those for which the expression for $f(x)$ is not defined, unless the problem specifies otherwise. An example of how a function might not be defined is in the case of $f(x) = \frac{1}{x+1}$, which is not defined when $x = -1$ (which would require dividing by zero). Therefore, in this case the domain would be all real numbers except $x = -1$.

If *y* is a function of *x*, then *x* is the *independent variable* and *y* is the *dependent variable*. This is because in many cases, the problem will start with some value of *x* and then see how *y* changes depending on this starting value.

In many cases, a function can be defined by giving an equation. For instance, $f(x) = x^2$ indicates that given a value for *x*, the output of *f* is found by squaring *x*.

Not all equations in *x* and *y* can be written in the form $y = f(x)$. An equation can be written in such a form if it satisfies the *vertical line test*: no vertical line meets the graph of the equation at more than a single point. In this case, *y* is said to be a *function of x*. If a vertical line meets the graph in two places, then this equation cannot be written in the form $y = f(x)$.

The graph of a function $f(x)$ is the graph of the equation $y = f(x)$. Thus, it is the set of all pairs (x, y) where $y = f(x)$. In other words, it is all pairs $(x, f(x))$. The *x*-intercepts are called the *zeros* of the function. The *y*-intercept is given by $f(0)$.

If, for a given function f, the only way to get $f(a) = f(b)$ is for $a = b$, then *f* is *one-to-one*. Often, even if a function is not one-to-one on its entire domain, it is one-to-one by considering a restricted portion of the domain.

A function $f(x) = k$ for some number *k* is called a *constant function*. The graph of a constant function is a horizontal line.

The function $f(x) = x$ is called the *identity function*. The graph of the identity function is the diagonal line pointing to the upper right at 45 degrees, $y = x$.

Given two functions, $f(x)$ and $g(x)$, new functions can be formed by adding, subtracting, multiplying, or dividing the functions. Any algebraic combination of the two functions can be performed, including one function being the exponent of the other. If there are expressions for *f* and *g*, then the result can be found by performing the desired operation between the expressions. So, if $f(x) = x^2$ and $g(x) = 3x$, then $f \times g(x) = x^2 \times 3x = 3x^3$.

Given two functions, $f(x)$ and $g(x)$, where the domain of *g* contains the range of *f*, the two functions can be combined together in a process called *composition*. The function—"*g* composed of *f*"—is written $(g \circ f)(x) = g(f(x))$. This requires the input of x into *f*, then taking that result and plugging it in to the function *g*.

If *f* is one-to-one, then there is also the option to find the function $f^{-1}(x)$, called the *inverse* of *f*. Algebraically, the inverse function can be found by writing y in place of $f(x)$, and then solving for *x*. The inverse function also makes this statement true: $f^{-1}(f(x)) = x$.

Computing the inverse of a function *f* entails the following procedure:

Given $f(x) = x^2$, with a domain of $x \geq 0$

$x = y^2$ is written down to find the inverse

The square root of both sides is determined to solve for *y*

Normally, this would mean $\pm\sqrt{x} = y$. However, the domain of *f* does not include the negative numbers, so the negative option needs to be eliminated.

The result is $y = \sqrt{x}$, so $f^{-1}(x) = \sqrt{x}$, with a domain of $x \geq 0$.

A function is called *monotone* if it is either always increasing or always decreasing. For example, the functions $f(x) = 3x$ and $f(x) = -x^5$ are monotone.

An *even function* looks the same when flipped over the *y*-axis: $f(x) = f(-x)$. The following image shows a graphic representation of an even function.

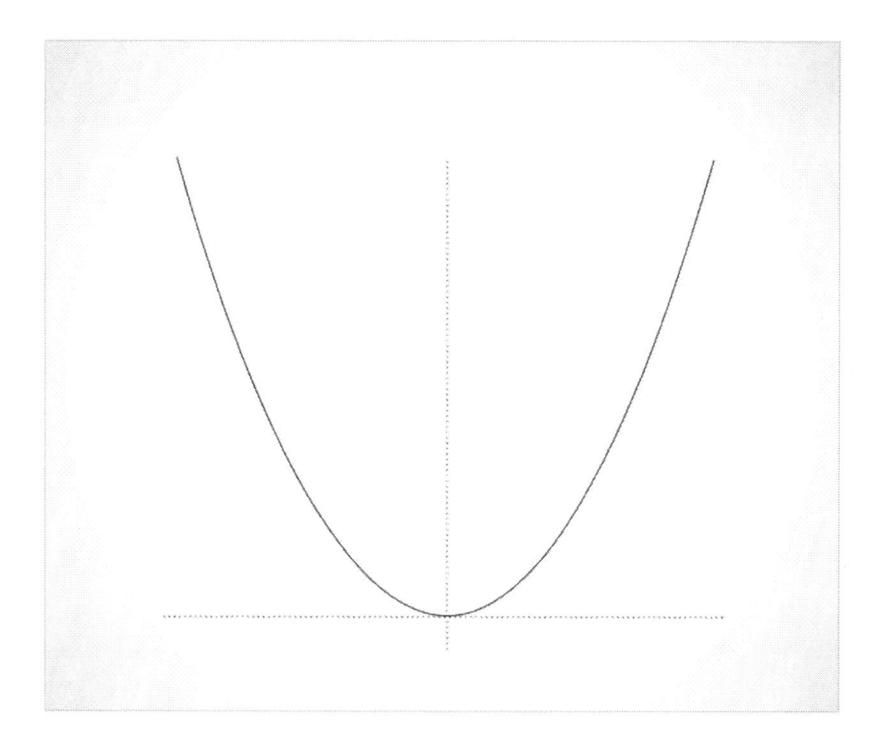

An *odd function* looks the same when flipped over the *y*-axis and then flipped over the *x*-axis: $f(x) = -f(-x)$. The following image shows an example of an odd function.

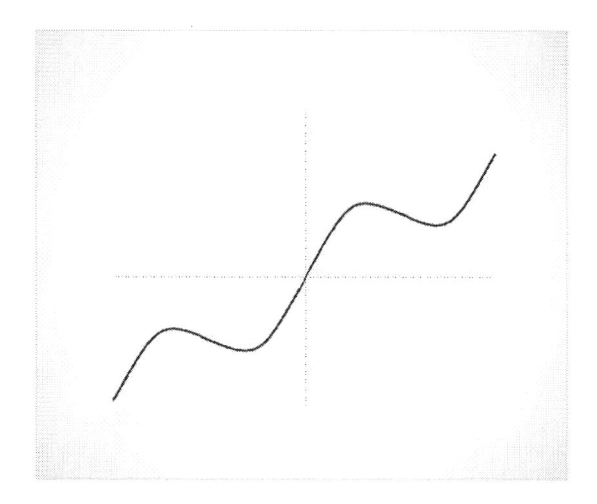

Evaluating Functions

To evaluate functions, plug in the given value everywhere the variable appears in the expression for the function. For example, find $g(-2)$ where $g(x) = 2x^2 - \frac{4}{x}$. To complete the problem, plug in -2 in the following way: $g(-2) = 2(-2)^2 - \frac{4}{-2} = 2 \times 4 + 2 = 8 + 2 = 10$.

Defining Linear Equations

A function is called *linear* if it can take the form of the equation $f(x) = ax + b$, or $y = ax + b$, for any two numbers *a* and *b*. A linear equation forms a straight line when graphed on the coordinate plane. An example of a linear function is shown below on the graph.

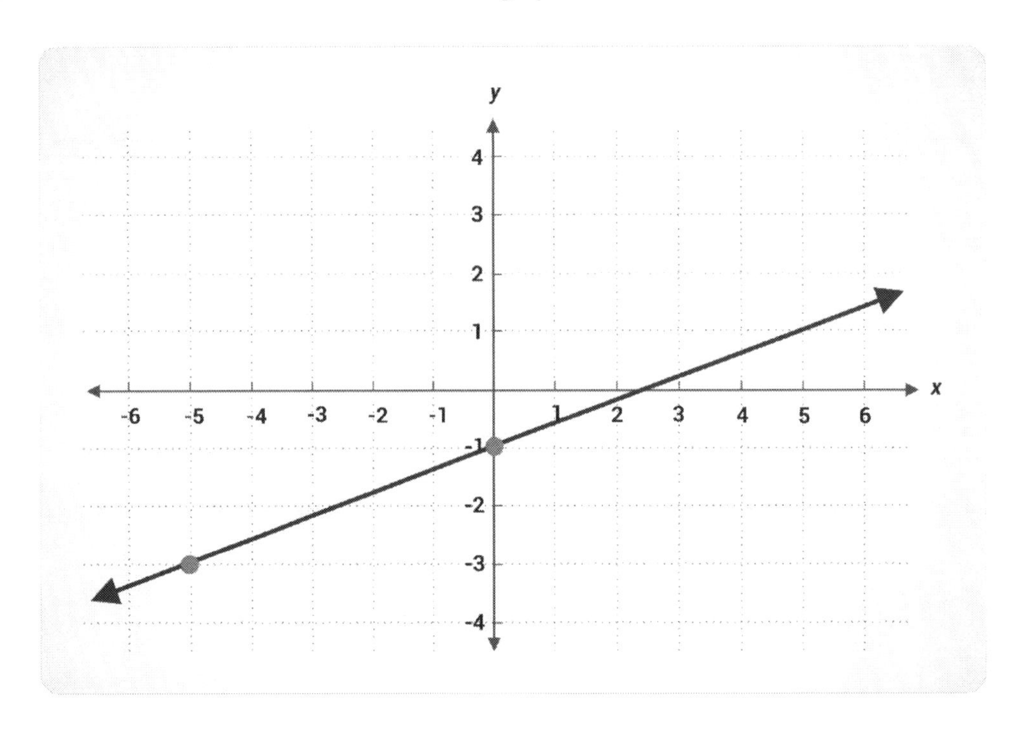

This is a graph of the following function: $y = \frac{2}{5}x - 1$. A table of values that satisfies this function is shown below.

X	Y
-5	-3
0	-1
5	1
10	3

These points can be found on the graph using the form (x,y). For more on graphing in the coordinate plane, refer to the Graphing section below.

Graphing Functions and Relations

To graph relations and functions, the Cartesian plane is used. This means to think of the plane as being given a grid of squares, with one direction being the *x*-axis and the other direction the *y*-axis. Generally, the independent variable is placed along the horizontal axis, and the dependent variable is placed along the vertical axis. Any point on the plane can be specified by saying how far to go along the *x*-axis and how far along the *y*-axis with a pair of numbers (x, y). Specific values for these pairs can be given names such as $C = (-1, 3)$. Negative values mean to move left or down; positive values mean to move right or up. The point where the axes cross one another is called the *origin*. The origin has coordinates $(0, 0)$ and is usually called *O* when given a specific label. An illustration of the Cartesian plane, along with graphs of $(2, 1)$ and $(-1, -1)$, are below.

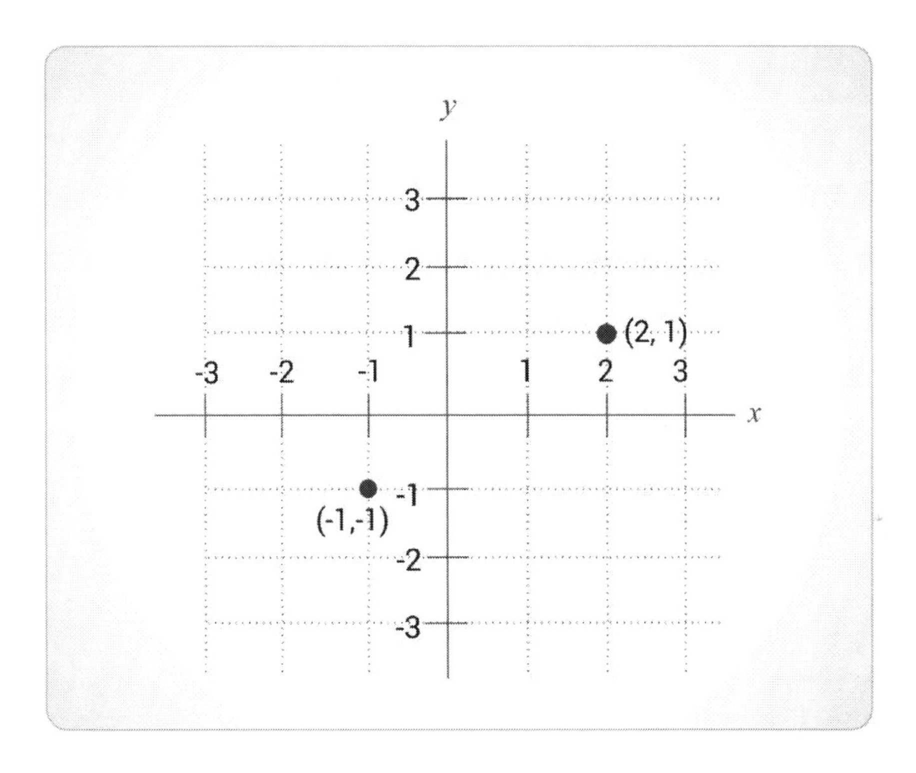

Relations also can be graphed by marking each point whose coordinates satisfy the relation. If the relation is a function, then there is only one value of *y* for any given value of x. This leads to the **vertical line test**: if a relation is graphed, then the relation is a function if any possible vertical line drawn anywhere along the graph would only touch the graph of the relation in no more than one place. Conversely, when graphing a function, then any possible vertical line drawn will not touch the graph of the function at any point or will touch the function at just one point. This test is made from the definition of a function, where each x-value must be mapped to one and only one y-value.

Forms of Linear Equations

When graphing a linear function, note that the ratio of the change of the *y* coordinate to the change in the *x* coordinate is constant between any two points on the resulting line, no matter which two points are chosen. In other words, in a pair of points on a line, (x_1, y_1) and (x_2, y_2), with $x_1 \neq x_2$ so that the two points are distinct, then the ratio $\frac{y_2 - y_1}{x_2 - x_1}$ will be the same, regardless of which particular pair of points are chosen. This ratio, $\frac{y_2-y_1}{x_2-x_1}$, is called the *slope* of the line and is frequently denoted with the letter m. If slope *m* is positive, then the line goes upward when moving to the right, while if slope *m* is negative, then the line goes downward when moving to the right. If the slope is 0, then the line is called *horizontal*, and the *y* coordinate is constant along the entire line. In lines where the *x* coordinate is constant along the entire line, *y* is not actually a function of *x*. For such lines, the slope is not defined. These lines are called *vertical* lines.

Linear functions may take forms other than $y = ax + b$. The most common forms of linear equations are explained below:

1. Standard Form: $Ax + By = C$, in which the slope is given by $m = \frac{-A}{B}$, and the *y*-intercept is given by $\frac{C}{B}$.
2. Slope-Intercept Form: $y = mx + b$, where the slope is *m* and the *y* intercept is *b*.
3. Point-Slope Form: $y - y_1 = m(x - x_1)$, where the slope is *m* and (x_1, y_1) is any point on the chosen line.
4. Two-Point Form: $\frac{y-y_1}{x-x_1} = \frac{y_2-y_1}{x_2-x_1}$, where (x_1, y_1) and (x_2, y_2) are any two distinct points on the chosen line. Note that the slope is given by $m = \frac{y_2-y_1}{x_2-x_1}$.
5. Intercept Form: $\frac{x}{x_1} + \frac{y}{y_1} = 1$, in which x_1 is the *x*-intercept and y_1 is the *y*-intercept.

These five ways to write linear equations are all useful in different circumstances. Depending on the given information, it may be easier to write one of the forms over another.

If $y = mx$, *y* is directly proportional to *x*. In this case, changing *x* by a factor changes *y* by that same factor. If $y = \frac{m}{x}$, *y* is inversely proportional to *x*. For example, if *x* is increased by a factor of 3, then *y* will be decreased by the same factor, 3.

Solving Linear Equations

Sometimes, rather than a situation where there's an equation such as $y = ax + b$ and finding *y* for some value of *x* is requested, the result is given and finding *x* is requested.

The key to solving any equation is to remember that from one true equation, another true equation can be found by adding, subtracting, multiplying, or dividing both sides by the same quantity. In this case, it's necessary to manipulate the equation so that one side only contains *x*. Then the other side will show what *x* is equal to.

For example, in solving $3x - 5 = 2$, adding 5 to each side results in $3x = 7$. Next, dividing both sides by 3 results in $x = \frac{7}{3}$. To ensure the value of x is correct, the number can be substituted into the original equation and solved to see if it makes a true statement. For example, $3(\frac{7}{3}) - 5 = 2$ can be simplified by cancelling out the two 3s. This yields $7 - 5 = 2$, which is a true statement.

Sometimes an equation may have more than one *x* term. For example, consider the following equation: $3x + 2 = x - 4$. Moving all of the *x* terms to one side by subtracting *x* from both sides results in $2x +$

$2 = -4$. Next, subtract 2 from both sides so that there is no constant term on the left side. This yields $2x = -6$. Finally, divide both sides by 2, which leaves $x = -3$.

Solving Linear Inequalities

Solving linear inequalities is very similar to solving equations, except for one rule: when multiplying or dividing an inequality by a negative number, the inequality symbol changes direction. Given the following inequality, solve for x: $-2x + 5 < 13$. The first step in solving this equation is to subtract 5 from both sides. This leaves the inequality: $-2x < 8$. The last step is to divide both sides by -2. By using the rule, the answer to the inequality is $x > -4$.

Since solutions to inequalities include more than one value, number lines are used many times to model the answer. For the previous example, the answer is modelled on the number line below. It shows that any number greater than -4, not including -4, satisfies the inequality.

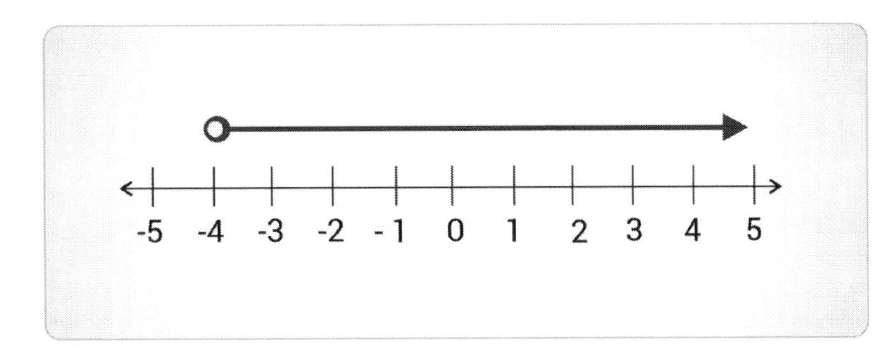

Linear Systems of Equations

A problem sometimes involves multiple variables and multiple equations. These are called *systems of equations*. In this case, try to manipulate them until an expression is found that provides the value of one of the variables. There are a couple of different approaches to this, and some of them can be used together in some cases. There are a few basic rules to keep in mind when solving systems of equations.

A single equation can be changed by doing the same operation to both sides, just as with one equation.

If one of the equations gives an expression for one of the variables in terms of other variables and constants, the expression can be substituted into the other equation, in place of the variable. This means the other equations will have one less variable in them.

If two equations are in the form of $a = b \text{ and } c = d$, then a new equation can be formed by adding the equations together, $a + c = b + d$, or subtracting the equations, $a - c = b - d$. This can eliminate one of the variables from an equation.

The general approach is to find a way to change one of the equations so that one variable is isolated, and then substitute that value (or expression) for the variable into the other equations.

The simplest case is a *linear system of two equations*, which has the form $ax + by = c, dx + ey = f$.

To solve linear systems of equations, use the same process to solve one equation in order to isolate one of the variables. Here's an example, using the linear system of equations:

$$2x - 3y = 2, 4x + 4y = 3$$

The first equation is multiplied on both sides by -2, which gives $-4x + 6y = -4$.

Adding this equation to the second equation will allow cancellation of the *x* term: $4x + 4y - 4x + 6y = 3 - 4$.

The result can be simplified to get $10y = -1$, which simplifies to $y = -\frac{1}{10}$.

The solution can be substituted into either of the original equations to find a value for *x*. Using the first equation, $2x - 3\left(-\frac{1}{10}\right) = 2$.

This simplifies to $2x + \frac{3}{10} = 2$, then to $2x = \frac{17}{10}$, and finally $x = \frac{17}{20}$.

The final solution is $x = \frac{17}{20}, y = -\frac{1}{10}$.

To check the validity of the answer, both solutions can be substituted into either original equation, which should result in a true statement.

An alternative way to solve this system would be to solve the first equation to get an expression for *y* in terms of *x*.

Subtracting $2x$ from both sides results in $-3y = 2 - 2x$.

Dividing both sides by -3 would be $y = \frac{2}{3}x - \frac{2}{3}$.

Then, this expression can be substituted into the second equation, getting $4x + 4\left(\frac{2}{3}x - \frac{2}{3}\right) = 3$.

This only involves the variable *x*, which can now be solved. Once the value for *x* is obtained, it can be substituted into either equation to solve for *y*.

There is one important issue to note here. If one of the equations in the system can be made to look identical to another equation, then it is *redundant*. The set of solutions will then be all pairs that satisfy the other equation.

For instance, in the system of equations, $2x - y = 1, -4x + 2y = -2$, the second equation can be made into the first equation by dividing both sides by -2. Thus, the solution set will be all pairs satisfying $2x - y = 1$, which simplifies to $y = 2x - 1$.

For a pair of linear equations, the simplest way to see if one equation is redundant is to rewrite each equation to the form $ax + by = c$. If one equation can be obtained from the other in this form by multiplying both sides by some constant, then the equations are redundant, and the answer to the system would be all real numbers.

It is also possible for the two equations to be *inconsistent*, which occurs when the two equations can be made into the form $ax + by = c, ax + by = d$, with *c* and *d* being different numbers. The two equations are inconsistent if, while trying to solve them, it is determined that an equation is false, such as $3 = 2$. This result shows that there are no solutions to that system of equations.

For linear systems of two equations with two variables, there will always be a single solution unless one of the two equations is redundant or the equations are inconsistent, in which case there are no solutions.

In summary, the three basic rules to keep in mind when solving systems of equations are the following.

1. Manipulate a set of equations by doing the same operation to both equations, just as is done when working with just one equation.
2. If one of the equations can be changed so that it expresses one variable in terms of the others, then that expression can be substituted into the other equations and the variable can be eliminated. This means the other equations will have one less variable in them. This is called the method of substitution.
3. If two equations of the form $a = b, c = d$ are included, then a new equation can be formed by adding the left sides and adding the right sides, $a + c = b + d$, or $a - c = b - d$. This enables the elimination of one of the variables from an equation. This is called the method of elimination.

The simplest case is the case of a *linear* system of equations. Although the equations may be written in more complicated forms, linear systems of equations with two variables can always be written in the form $ax + by = c, dx + ey = f$. The two basic approaches to solving these systems are substitution and elimination.

Consider the system $3x - y = 2 \; and \; 2x + 2y = 3$. This can be solved in two ways.

1. By substitution: start by solving the first equation for y. First, subtract $3x$ from both sides to obtain $-y = 2 - 3x$. Next, divide both sides by -1, to obtain $y = 3x - 2$. Then substitute this value for y into the second equation. This yields $2x + 2(3x - 2) = 3$. This can be simplified to $2x + 6x - 4 = 3$, or $8x = 7$, which means $x = \frac{7}{8}$. By plugging in this value for x into $y = 3x - 2$, the result is $y = 3\left(\frac{7}{8}\right) - 2 = \frac{21}{8} - \frac{16}{8} = \frac{5}{8}$. So, this results in $x = \frac{7}{8}, y = \frac{5}{8}$.
2. By elimination: first, multiply the first equation by 2. This results in $-2y$, which could cancel out the +$2y$ in the second equation. Multiplying both sides of the first equation by 2 gives results in $2(3x - y) = 2(2)$, or $6x - 2y = 4$. Adding the left sides and the right sides of the two equations and setting the results equal to one another results in $(6x + 2x) + (-2y + 2y) = 4 + 3$. This simplifies to $8x = 7$, so again $x = \frac{7}{8}$. Plug this back into either of the original equations and the result is $3\left(\frac{7}{8}\right) - y = 2$ or $y = 3\left(\frac{7}{8}\right) - 2 = \frac{21}{8} - \frac{16}{8} = \frac{5}{8}$. This again yields $x = \frac{7}{8}, y = \frac{5}{8}$.

As this shows, both methods will give the same answer. However, one method is sometimes preferred over another simply because of the amount of work required. To check the answer, the values can be substituted into the given system to make sure they form two true statements.

Algebraic Expressions and Equations

Algebraic expressions look similar to equations, but they do not include the equal sign. Algebraic expressions are comprised of numbers, variables, and mathematical operations. Some examples of algebraic expressions are $8x + 7y - 12z$, $3a^2$, and $5x^3 - 4y^4$.

Algebraic expressions and equations can be used to represent real-life situations and model the behavior of different variables. For example, $2x + 5$ could represent the cost to play games at an arcade. In this case, 5 represents the price of admission to the arcade, and 2 represents the cost of each game played. To calculate the total cost, use the number of games played for x, multiply it by 2, and add 5.

Evaluation of Simple Formulas and Expressions

To evaluate simple formulas and expressions, the first step is to substitute the given values in for the variable(s). Then, the order of operations is used to simplify.

Example 1

Evaluate $\frac{1}{2}x^2 - 3, x = 4$.

The first step is to substitute in 4 for *x* in the expression: $\frac{1}{2}(4)^2 - 3$.

Then, the order of operations is used to simplify.

The exponent comes first, $\frac{1}{2}(16) - 3$, then the multiplication 8 – 3, and then, after subtraction, the solution is 5.

Example 2

Evaluate $4|5 - x| + 2y, x = 4, y = -3$.

The first step is to substitute 4 in for *x* and -3 in for *y* in the expression: $4|5 - 4| + 2(-3)$.

Then, the absolute value expression is simplified, which is $|5 - 4| = |1| = 1$.

The expression is $4(1) + 2(-3)$ which can be simplified using the order of operations.

First is the multiplication, 4 + (-6); then addition yields an answer of -2.

Example 3

Find the perimeter of a rectangle with a length of 6 inches and a width of 9 inches.

The first step is substituting in 6 for the length and 9 for the width in the perimeter of a rectangle formula, $P = 2(6) + 2(9)$.

Then, the order of operations is used to simplify.

First is multiplication (resulting in 12 + 18) and then addition for a solution of 30 inches.

Simplifying Algebraic Fractions

A *rational expression* is a fraction with a polynomial in the numerator and denominator. The denominator polynomial cannot be zero. An example of a rational expression is $\frac{3x^4-2}{-x+1}$. The same rules for working with addition, subtraction, multiplication, and division with rational expressions apply as when working with regular fractions.

The first step is to find a common denominator when adding or subtracting. This can be done just as with regular fractions. For example, if $\frac{a}{b}+\frac{c}{d}$, then a common denominator can be found by multiplying to find the following fractions: $\frac{ad}{bd}, \frac{cb}{db}$.

A *complex fraction* is a fraction in which the numerator and denominator are themselves fractions, of the form $\frac{(\frac{a}{b})}{(\frac{c}{d})}$. These can be simplified by following the usual rules for the order of operations, or by remembering that dividing one fraction by another is the same as multiplying by the reciprocal of the divisor. This means that any complex fraction can be rewritten using the following form: $\frac{(\frac{a}{b})}{(\frac{c}{d})} = \frac{a}{b} \times \frac{d}{c}$.

The following problem is an example of solving a complex fraction:

$$\frac{(\frac{5}{4})}{(\frac{3}{8})} = \frac{5}{4} \times \frac{8}{3} = \frac{40}{12} = \frac{10}{3}$$

Word Problems and Applications

In word problems, multiple quantities are often provided with a request to find some kind of relation between them. This often will mean that one variable (the dependent variable whose value needs to be found) can be written as a function of another variable (the independent variable whose value can be figured from the given information). The usual procedure for solving these problems is to start by giving each quantity in the problem a variable, and then figuring the relationship between these variables.

For example, suppose a car gets 25 miles per gallon. How far will the car travel if it uses 2.4 gallons of fuel? In this case, *y* would be the distance the car has traveled in miles, and *x* would be the amount of fuel burned in gallons (2.4). Then the relationship between these variables can be written as an algebraic equation, $y = 25x$. In this case, the equation is $y = 25 \times 2.4 = 60$, so the car has traveled 60 miles.

Some word problems require more than just one simple equation to be written and solved. Consider the following situations and the linear equations used to model them.

Suppose Margaret is 2 miles to the east of John at noon. Margaret walks to the east at 3 miles per hour. How far apart will they be at 3 p.m.? To solve this, *x* would represent the time in hours past noon, and *y* would represent the distance between Margaret and John. Now, noon corresponds to the equation where *x* is 0, so the *y* intercept is going to be 2. It's also known that the slope will be the rate at which the distance is changing, which is 3 miles per hour. This means that the slope will be 3 (be careful at this point: if units were used, other than miles and hours, for *x* and *y* variables, a conversion of the given information to the appropriate units would be required first). The simplest way to write an equation given the y-intercept, and the slope is the Slope-Intercept form, is $y = mx + b$. Recall that m here is the slope, and b is the y intercept. So, $m = 3$ and $b = 2$. Therefore, the equation will be $y = 3x + 2$. The word problem asks how far to the east Margaret will be from John at 3 p.m., which means when x is 3. So, substitute $x = 3$ into this equation to obtain $y = 3 \times 3 + 2 = 9 + 2 = 11$. Therefore, she will be 11 miles to the east of him at 3 p.m.

For another example, suppose that a box with 4 cans in it weighs 6 lbs., while a box with 8 cans in it weighs 12 lbs. Find out how much a single can weighs. To do this, let x denote the number of cans in the box, and y denote the weight of the box with the cans in lbs. This line touches two pairs: $(4, 6)$ and (8, 12). A formula for this relation could be written using the two-point form, with $x_1 = 4, y_1 = 6, x_2 = 8, y_2 = 12$. This would yield $\frac{y-6}{x-4} = \frac{12-6}{8-4}$, or $\frac{y-6}{x-4} = \frac{6}{4} = \frac{3}{2}$. However, only the slope is needed to solve this problem, since the slope will be the weight of a single can. From the computation, the slope is $\frac{3}{2}$. Therefore, each can weighs $\frac{3}{2}$ lb.

In general, when solving word problems presented in an algebraic context, there is a four-step process in problem-solving that can be used as a guide:

1. Understand the problem and determine the unknown information.
2. Translate the verbal problem into an algebraic equation.
3. Solve the equation by using inverse operations.
4. Check the work and answer the given question.

Example

Three times the sum of a number plus 4 equals the number plus 8. What is the number?

The first step is to determine the unknown, which is the number, or x.

The second step is to translate the problem into the equation, which is $3(x + 4) = x + 8$.

The equation can be solved as follows:

$3x + 12 = x + 8$	Apply the distributive property
$3x = x - 4$	Subtract 12 from both sides of the equation
$2x = -4$	Subtract x from both sides of the equation
$x = -2$	Divide both sides of the equation by 2

The final step is checking the solution. Plugging the value for x back into the equation yields the following problem: $3(-2) + 12 = -2 + 8$. Using the order of operations shows that a true statement is made: $6 = 6$.

Translation of Written Phases into Algebraic Expressions

An *algebraic expression* contains one or more operations and one or more variables. To convert written phrases into algebraic expression, there are some key terms to recognize:

- Key terms with addition are *sum*, *increase*, *plus*, *add*, *more than*, and *total.*
- Key terms with subtraction are *difference*, *decrease*, *minus*, *subtract*, and *less than*.
- Key terms with multiplication are *product, times*, and *multiplied*.
- Key terms with division are *quotient, divided,* and *ratio*.
- Key terms with exponent are *squares, cubed,* and *raised to a power*.

To write a phrase as an algebraic expression, it's necessary to identify the unknown(s) where variables will be used and the words for the correct operation.

Example 1

Write an expression for three times the sum of twice the number *n* plus five.

Three times means *3 x*, twice a number and five means *2n + 5*, and the final expression is *3(2n + 5)*.

Example 2

Write an expression for the total price of $2 per pound for grapes and $3 per pound for strawberries.

The total means the sum. The price for grapes is *2g*, and the price for strawberries is *3s*. The expression is *2g + 3s*.

Intermediate Algebra and Functions

Polynomials

An expression of the form ax^n, where *n* is a non-negative integer, is called a *monomial* because it contains one term. A sum of monomials is called a *polynomial*. For example, $-4x^3 + x$ is a polynomial, while $5x^7$ is a monomial. A function equal to a polynomial is called a *polynomial function*.

The monomials in a polynomial are also called the *terms* of the polynomial.

The constants that precede the variables are called *coefficients*.

The highest value of the exponent of *x* in a polynomial is called the *degree* of the polynomial. So, $-4x^3 + x$ has a degree of 3, while $-2x^5 + x^3 + 4x + 1$ has a degree of 5. When multiplying polynomials, the degree of the result will be the sum of the degrees of the two polynomials being multiplied.

To add polynomials, add the coefficients of like powers of *x*. For example:

$$(-2x^5 + x^3 + 4x + 1) + (-4x^3 + x)$$
$$-2x^5 + (1 - 4)x^3 + (4 + 1)x + 1$$
$$-2x^5 - 3x^3 + 5x + 1$$

Likewise, subtraction of polynomials is performed by subtracting coefficients of like powers of x. So,

$$(-2x^5 + x^3 + 4x + 1) - (-4x^3 + x)$$
$$-2x^5 + (1 + 4)x^3 + (4 - 1)x + 1$$
$$-2x^5 + 5x^3 + 3x + 1$$

To multiply two polynomials, multiply each term of the first polynomial by each term of the second polynomial and add the results. For example:

$$(4x^2 + x)(-x^3 + x)$$

$$4x^2(-x^3) + 4x^2(x) + x(-x^3) + x(x)$$

$$-4x^5 + 4x^3 - x^4 + x^2$$

In the case where each polynomial has two terms, like in this example, some students find it helpful to remember this as multiplying the First terms, then the Outer terms, then the Inner terms, and finally the Last terms, with the mnemonic FOIL. For longer polynomials, the multiplication process is the same, but there will be, of course, more terms, and there is no common mnemonic to remember each combination.

To divide one polynomial by another, the procedure is similar to long division. At each step, one needs to figure out how to get the term of the dividend with the highest exponent as a multiple of the divisor. The divisor is multiplied by the multiple to get that term, which goes in the quotient. Then, the product of this term is subtracted with the dividend from the divisor and repeat the process. This sounds rather abstract, so it may be easiest to see the procedure by describing it while looking at an example.

Example

$(4x^3 + x^2 - x + 4) \div (2x - 1)$

The first step is to cancel out the highest term in the first polynomial.

To get $4x^3$ from the second polynomial, multiply by $2x^2$.

The first term for the quotient is going to be $2x^2$.

The result of $2x^2(2x - 1)$ is $4x^3 - 2x^2$. Subtract this from the first polynomial.

The result is $(-x^2 - x + 4) \div (2x - 1)$.

The procedure is repeated: to cancel the $-x^2$ term, then multiply $(2x - 1)$ by $-\frac{1}{2}x$.

Adding this to the quotient, the quotient becomes $2x^2 - \frac{1}{2}x$.

The dividend is changed by subtracting $-\frac{1}{2}x(2x - 1)$ from it to obtain $(-\frac{3}{2}x + 4) \div (2x - 1)$.

To get $-\frac{3}{2}x$ needs to be multiplied by $-\frac{3}{4}$.

The quotient, therefore, becomes $2x^2 - \frac{1}{2}x - \frac{3}{4}$.

The remaining part is $4.75 \div (2x - 1)$.

There is no monomial to multiply to cancel this constant term, since the divisor now has a higher power than the dividend.

The final answer is the quotient plus the remainder divided by $(2x - 1)$

$$2x^2 - \frac{1}{2}x - \frac{3}{4} + \frac{4.75}{2x - 1}$$

Factors for polynomials are similar to factors for integers—they are numbers, variables, or polynomials that, when multiplied together, give a product equal to the polynomial in question. One polynomial is a factor of a second polynomial if the second polynomial can be obtained from the first by multiplying by a third polynomial.

$6x^6 + 13x^4 + 6x^2$ can be obtained by multiplying together $(3x^4 + 2x^2)(2x^2 + 3)$. This means $2x^2 + 3$ and $3x^4 + 2x^2$ are factors of $6x^6 + 13x^4 + 6x^2$.

In general, finding the factors of a polynomial can be tricky. However, there are a few types of polynomials that can be factored in a straightforward way.

If a certain monomial is in each term of a polynomial, it can be factored out. There are several common forms polynomials take, which if you recognize, you can solve. The first example is a perfect square trinomial. To factor this polynomial, first expand the middle term of the expression:

$$x^2 + 2xy + y^2$$

$$x^2 + xy + xy + y^2$$

Factor out a common term in each half of the expression (in this case x from the left and y from the right):

$$x(x + y) + y(x + y)$$

Then the same can be done again, treating $(x + y)$ as the common factor:

$$(x + y)(x + y) = (x + y)^2$$

Therefore, the formula for this polynomial is:

$$x^2 + 2xy + y^2 = (x + y)^2$$

Next is another example of a perfect square trinomial. The process is the similar, but notice the difference in sign:

$$x^2 - 2xy + y^2$$

$$x^2 - xy - xy + y^2$$

Factor out the common term on each side:

$$x(x - y) - y(x - y)$$

Factoring out the common term again:

$$(x-y)(x-y)=(x-y)^2$$

Thus:

$$x^2-2xy+y^2=(x-y)^2$$

The next is known as a difference of squares. This process is effectively the reverse of binomial multiplication:

$$x^2-y^2$$

$$x^2-xy+xy-y^2$$

$$x(x-y)+y(x-y)$$

$$(x+y)(x-y)$$

Therefore:

$$x^2-y^2=(x+y)(x-y)$$

The following two polynomials are known as the sum or difference of cubes. These are special polynomials that take the form of x^3+y^3 or x^3-y^3. The following formula factors the sum of cubes:

$$x^3+y^3=(x+y)(x^2-xy+y^2)$$

Next is the difference of cubes, but note the change in sign. The formulas for both are similar, but the order of signs for factoring the sum or difference of cubes can be remembered by using the acronym SOAP, which stands for "same, opposite, always positive." The first sign is the same as the sign in the first expression, the second is opposite, and the third is always positive. The next formula factors the difference of cubes:

$$x^3-y^3=(x-y)(x^2+xy+y^2)$$

The following two examples are expansions of cubed binomials. Similarly, these polynomials always follow a pattern:

$$x^3+3x^2y+3xy^2+y^3=(x+y)^3$$

$$x^3-3x^2y+3xy^2-y^3=(x-y)^3$$

These rules can be used in many combinations with one another. For example, the expression $3x^3-24$ has a common factor of 3, which becomes:

$$3(x^3-8)$$

A difference of cubes still remains which can then be factored out:

$$3(x-2)(x^2+2x+4)$$

There are no other terms to be pulled out, so this expression is completely factored.

When factoring polynomials, a good strategy is to multiply the factors to check the result. Let's try another example:

$$4x^3 + 16x^2$$

Both sides of the expression can be divided by 4, and both contain x^2, because $4x^3$ can be thought of as $4x^2(x)$, so the common term can simply be factored out:

$$4x^2(x + 4)$$

It sometimes can be necessary to rewrite the polynomial in some clever way before applying the above rules. Consider the problem of factoring $x^4 - 1$. This does not immediately look like any of the previous polynomials. However, it's possible to think of this polynomial as $x^4 - 1 = (x^2)^2 - (1^2)^2$, and now it can be treated as a difference of squares to simplify this:

$$(x^2)^2 - (1^2)^2$$

$$(x^2)^2 - x^2 1^2 + x^2 1^2 - (1^2)^2$$

$$x^2(x^2 - 1^2) + 1^2(x^2 - 1^2)$$

$$(x^2 + 1^2)(x^2 - 1^2)$$

$$(x^2 + 1)(x^2 - 1)$$

Expanding Polynomials

Some polynomials may need to be expanded to identify the final solution—*polynomial expansion* means that parenthetical polynomials are multiplied out so that the parentheses no longer exist. The polynomials will be in the form $(a + b)^n$ where n is a whole number greater than 2. The expression can be simplified using the *distributive property*, which states that a number, variable, or polynomial that is multiplied by a polynomial in parentheses should be multiplied by each item in the parenthetical polynomial. Here's one example:

$$(a + b)^2 = (a + b)(a + b) = a^2 + ab + ab + b^2 = a^2 + 2ab + b^2$$

Here's another example to consider:

$$(a + b)^3$$

$$(a + b)(a + b)(a + b)$$

$$(a^2 + ab + ab + b^2)(a + b)$$

$$(a^2 + 2ab + b^2)(a + b)$$

$$a^3 + 2a^2b + ab^2 + a^2b + 2ab^2 + b^3$$

$$a^3 + 3a^2b + 3ab^2 + b^3$$

Quadratic Functions

A polynomial of degree 2 is called *quadratic*. Every quadratic function can be written in the form $ax^2 + bx + c$. The graph of a quadratic function, $y = ax^2 + bx + c$, is called a *parabola*. Parabolas are vaguely U-shaped.

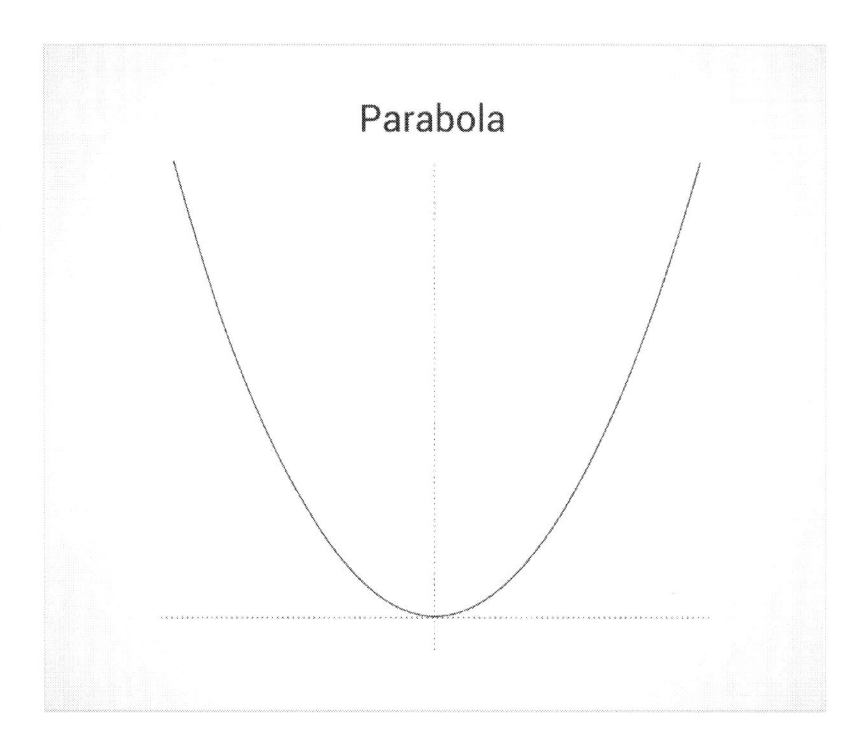

Whether the parabola opens upward or downward depends on the sign of a. If a is positive, then the parabola will open upward. If a is negative, then the parabola will open downward. The value of a will also affect how wide the parabola is. If the absolute value of a is large, then the parabola will be fairly skinny. If the absolute value of a is small, then the parabola will be quite wide.

Changes to the value of b affect the parabola in different ways, depending on the sign of a. For positive values of a, increasing b will move the parabola to the left, and decreasing b will move the parabola to the right. On the other hand, if a is negative, the effects will be the opposite: increasing b will move the parabola to the right, while decreasing b will move the parabola to the left.

Changes to the value of c move the parabola vertically. The larger that c is, the higher the parabola gets. This does not depend on the value of a.

The quantity $D = b^2 - 4ac$ is called the *discriminant* of the parabola. When the discriminant is positive, then the parabola has two real zeros, or x intercepts. However, if the discriminant is negative, then there are no real zeros, and the parabola will not cross the x-axis. The highest or lowest point of the parabola is called the *vertex*. If the discriminant is zero, then the parabola's highest or lowest point is on the x-axis, and it will have a single real zero. The x-coordinate of the vertex can be found using the equation $x = -\frac{b}{2a}$. Plug this x-value into the equation and find the y-coordinate.

A quadratic equation is often used to model the path of an object thrown into the air. The x-value can represent the time in the air, while the y-value can represent the height of the object. In this case, the maximum height of the object would be the y-value found when the x-value is $-\frac{b}{2a}$.

Solving Quadratic Equations

A *quadratic equation* is an equation in the form $ax^2 + bx + c = 0$. There are several methods to solve such equations. The easiest method will depend on the quadratic equation in question.

Solving Quadratic Equations by Factoring

It sometimes is possible to solve quadratic equations by manually *factoring* them. This means rewriting them in the form $(x + A)(x + B) = 0$. If this is done, then they can be solved by remembering that when $ab = 0$, either *a* or *b* must be equal to zero. Therefore, to have $(x + A)(x + B) = 0$, $(x + A) = 0$ or $(x + B) = 0$ is needed. These equations have the solutions $x = -A$ and $x = -B$, respectively.

In order to factor a quadratic equation, note that $(x + A)(x + B) = x^2 + (A + B)x + AB$. So, if an equation is in the form $x^2 + bx + c$, two numbers, *A* and *B,* need to be found that will add up to give us *b*, and multiply together to give *c*.

If the equation takes the form $ax^2 - b = 0$, then it can be solved by adding *b* to both sides and dividing by *a* to get $x^2 = \frac{b}{a}$.

Using the sixth rule discussed for evaluating rational roots and exponents, the solution is $x = \pm\sqrt{\frac{b}{a}}$. Note that this is actually two separate solutions, unless *b* happens to be 0.

If a quadratic equation has no constant—so that it takes the form $ax^2 + bx = 0$—then the *x* can be factored out to get $x(ax + b) = 0$. Then, the solutions are $x = 0$, together with the solutions to $ax + b = 0$. Both factors x *and* $(ax + b)$ can be set equal to zero to solve for x because one of those values must be zero for their product to equal zero. For an equation $ab = 0$ to be true, either $a = 0$, or $b = 0$.

A given quadratic equation $x^2 + bx + c$ can be factored into $(x + A)(x + B)$, where $A + B = b$, and $AB = c$. Finding the values of A and B can take time, but such a pair of numbers can be found by guessing and checking. Looking at the positive and negative factors for *c* offers a good starting point.

For example, in $x^2 - 5x + 6$, the factors of 6 are 1, 2, and 3. Now,$(-2)(-3) = 6$, and $-2 - 3 = -5$. In general, however, this may not work, in which case another approach may need to be used.

A quadratic equation of the form $x^2 + 2xb + b^2 = 0$ can be factored into $(x + b)^2 = 0$. Similarly, $x^2 - 2xy + y^2 = 0$ factors into $(x - y)^2 = 0$.

In general, the constant term may not be the right value to be factored this way. A more general method for solving these quadratic equations must then be found. The following two methods will work in any situation.

Completing the Square

The first method is called *completing the square*. The idea here is that in any equation $x^2 + 2xb + c = 0$, something could be added to both sides of the equation to get the left side to look like $x^2 + 2xb + b^2$, meaning it could be factored into $(x + b)^2 = 0$.

Example

$x^2 + 6x - 1 = 0$

The left-hand side could be factored if the constant were equal to 9, , since $x^2 + 6x + 9 = (x + 3)^2$.

To get a constant of 9 on the left, 10 needs to be added to both sides.

That changes the equation to $x^2 + 6x + 9 = 10$.

Factoring the left gives $(x + 3)^2 = 10$.

Then, the square root of both sides can be taken (remembering that this introduces a $\pm$): $x + 3 = \pm\sqrt{10}$.

Finally, 3 is subtracted from both sides to get two solutions: $x = -3 \pm \sqrt{10}$.

The Quadratic Formula

The first method of completing the square can be used in finding the second method, the quadratic formula. It can be used to solve any quadratic equation. This formula may be the longest method for solving quadratic equations and is commonly used as a last resort after other methods are ruled out.

It can be helpful in memorizing the formula to see where it comes from, so here are the steps involved.

The most general form for a quadratic equation is $ax^2 + bx + c = 0$.

First, dividing both sides by a leaves us with $x^2 + \frac{b}{a}x + \frac{c}{a} = 0$.

To complete the square on the left-hand side, c/a can be subtracted on both sides to get $x^2 + \frac{b}{a}x = -\frac{c}{a}$.

$(\frac{b}{2a})^2$ is then added to both sides.

This gives $x^2 + \frac{b}{a}x + (\frac{b}{2a})^2 = (\frac{b}{2a})^2 - \frac{c}{a}$.

The left can now be factored and the right-hand side simplified to give $(x + \frac{b}{2a})^2 = \frac{b^2 - 4ac}{4a}$.

Taking the square roots gives:

$$x + \frac{b}{2a} = \pm\frac{\sqrt{b^2 - 4ac}}{2a}$$

Solving for x yields the quadratic formula:

$$x = \frac{-b \pm \sqrt{b^2 - 4ac}}{2a}$$

It isn't necessary to remember how to get this formula, but memorizing the formula itself is the goal.

If an equation involves taking a root, then the first step is to move the root to one side of the equation and everything else to the other side. That way, both sides can be raised to the index of the radical in order to remove it, and solving the equation can continue.

Algebraic Functions

A function is called *algebraic* if it is built up from polynomials by adding, subtracting, multiplying, dividing, and taking radicals. This means that, for example, the variable can never appear in an exponent. Thus, polynomials and rational functions are algebraic, but exponential functions are not algebraic. It turns out that logarithms and trigonometric functions are not algebraic either.

A function of the form $f(x) = a_n x^n + a_{n-1}x^{n-1} + a_{n-2}x^{n-2} + \cdots + a_1 x + a_0$ is called a *polynomial function*. The value of *n* is called the *degree* of the polynomial. In the case where $n = 1$, it is called a *linear function*. In the case where $n = 2$, it is called a *quadratic function*. In the case where $n = 3$, it is called a *cubic function*.

When *n* is even, the polynomial is called *even*, and not all real numbers will be in its range. When *n* is odd, the polynomial is called *odd*, and the range includes all real numbers.

The graph of a quadratic function $f(x) = ax^2 + bx + c$ will be a parabola. To see whether or not the parabola opens up or down, it's necessary to check the coefficient of x^2, which is the value a.

If the coefficient is positive, then the parabola opens upward. If the coefficient is negative, then the parabola opens downward.

The quantity $D = b^2 - 4ac$ is called the *discriminant* of the parabola. If the discriminant is positive, then the parabola has two real zeros. If the discriminant is negative, then it has no real zeros.

If the discriminant is zero, then the parabola's highest or lowest point is on the *x*-axis, and it has a single real zero.

The highest or lowest point of the parabola is called the *vertex*. The coordinates of the vertex are given by the point $(-\frac{b}{2a}, -\frac{D}{4a})$. The roots of a quadratic function can be found with the quadratic formula, which is:

$$x = \frac{-b \pm \sqrt{b^2 - 4ac}}{2a}$$

A *rational function* is a function $f(x) = \frac{p(x)}{q(x)}$, where *p* and *q* are both polynomials. The domain of *f* will be all real numbers except the (real) roots of *q*.

At these roots, the graph of *f* will have a *vertical asymptote*, unless they are also roots of *p*. Here is an example to consider:

$$p(x) = p_n x^n + p_{n-1}x^{n-1} + p_{n-2}x^{n-2} + \cdots + p_1 x + p_0$$

$$q(x) = q_m x^m + q_{m-1}x^{m-1} + q_{m-2}x^{m-2} + \cdots + q_1 x + q_0$$

When the degree of *p* is less than the degree of *q*, there will be a horizontal asymptote of $y = 0$. If *p* and *q* have the same degree, there will be a horizontal asymptote of $y = \frac{p_n}{q_n}$. If the degree of *p* is exactly one greater than the degree of *q*, then *f* will have an oblique asymptote along the line $y = \frac{p_n}{q_{n-1}}x + \frac{p_{n-1}}{q_{n-1}}$.©

Graphs of Algebraic Functions

A graph can shift in many ways. To shift it horizontally, a constant can be added to all the *x* variables. Replacing x with $(x + a)$ will shift the graph to the left by *a*. If *a* is negative, this shifts the graph to the right. Similarly, vertical shifts occur by adding a constant to each of the *y* variables. Replacing y by $(y + a)$ will shift the graph up by *a*. If *a* is negative, then it shifts the graph down.

A graph can also stretch and shrink the graph in the horizontal and vertical directions. To stretch by a (positive) factor of *k* horizontally, all instances of x are replaced with $\frac{x}{k}$. To stretch vertically by *k*, all instances of y are replaced with $\frac{y}{k}$.

The graph can be reflected over the *y*-axis by replacing all instances of x with $(-x)$. The graph can also be reflected over the *x*-axis by replacing all instances of y with $(-y)$.

Rational Expressions

A rational expression is a fraction where the numerator and denominator are both polynomials. Some examples of rational expressions include the following: $\frac{4x^3y^5}{3z^4}$, $\frac{4x^3+3x}{x^2}$, and $\frac{x^2+7x+10}{x+2}$. Since these refer to expressions and not equations, they can be simplified but not solved. Using the rules in the previous Exponents and Roots sections, some rational expressions with monomials can be simplified. Other rational expressions such as the last example, $\frac{x^2+7x+10}{x+2}$, take more steps to be simplified. First, the polynomial on top can be factored from $x^2 + 7x + 10$ into $(x + 5)(x + 2)$. Then the common factors can be canceled and the expression can be simplified to $(x + 5)$.

Consider this problem as an example of using rational expressions. Reggie wants to lay sod in his rectangular backyard. The length of the yard is given by the expression $4x + 2$ and the width is unknown. The area of the yard is $20x + 10$. Reggie needs to find the width of the yard. Knowing that the area of a rectangle is length multiplied by width, an expression can be written to find the width: $\frac{20x+10}{4x+2}$, area divided by length. Simplifying this expression by factoring out 10 on the top and 2 on the bottom leads to this expression: $\frac{10(2x+1)}{2(2x+1)}$. By cancelling out the $2x + 1$, that results in $\frac{10}{2} = 5$. The width of the yard is found to be 5 by simplifying a rational expression.

Rational Equations

A *rational equation* can be as simple as an equation with a ratio of polynomials, $\frac{p(x)}{q(x)}$, set equal to a value, where $p(x)$ and $q(x)$ are both polynomials. Notice that a rational equation has an equal sign, which is different from expressions. This leads to solutions, or numbers that make the equation true.

It is possible to solve rational equations by trying to get all of the *x* terms out of the denominator and then isolate them on one side of the equation. For example, to solve the equation $\frac{3x+2}{2x+3} = 4$, start by multiplying both sides by $(2x + 3)$. This will cancel on the left side to yield $3x + 2 = 4(2x + 3)$, then $3x + 2 = 8x + 12$. Now, subtract $8x$ from both sides, which yields $-5x + 2 = 12$. Subtracting 2 from both sides results in $-5x = 10$. Finally, divide both sides by -5 to obtain $x = -2$.

Sometimes, when solving rational equations, it can be easier to try to simplify the rational expression by factoring the numerator and denominator first, then cancelling out common factors. For example, to solve $\frac{2x^2-8x+6}{x^2-3x+2} = 1$, start by factoring $2x^2 - 8x + 6 = 2(x^2 - 4x + 3) = 2(x - 1)(x - 3)$. Then, factor $x^2 - 3x + 2$ into $(x - 1)(x - 2)$. This turns the original equation into $\frac{2(x-1)(x-3)}{(x-1)(x-2)} = 1$. The common factor of $(x - 1)$ can be canceled, leaving us with $\frac{2(x-3)}{x-2} = 1$. Now the same method used in the previous example can be followed. Multiplying both sides by $x - 2$ and performing the multiplication on the left yields $2x - 6 = x - 2$, which can be simplified to $x = 4$.

Rational Functions

A rational function is similar to an equation, but it includes two variables. In general, a rational function is in the form: $f(x) = \frac{p(x)}{q(x)}$, where $p(x)$ and $q(x)$ are polynomials. Refer to the previous Functions section for a more detailed definition of functions. Rational functions are defined everywhere except where the denominator is equal to zero. When the denominator is equal to zero, this indicates either a hole in the graph or an asymptote. An asymptote can be either vertical, horizontal, or slant. A hole occurs when both the numerator and denominator are equal to 0 for a given value of *x*. A rational function can have at most one vertical asymptote and one horizontal or slant asymptote. An asymptote is a line such that the distance between the curve and the line tends toward 0, but never reaches it, as the line heads toward infinity. Examples of these types of functions are shown below. The first graph shows a rational function with a vertical asymptote at x = 0. This can be found by setting the denominator equal to 0. In this case it is just x = 0.

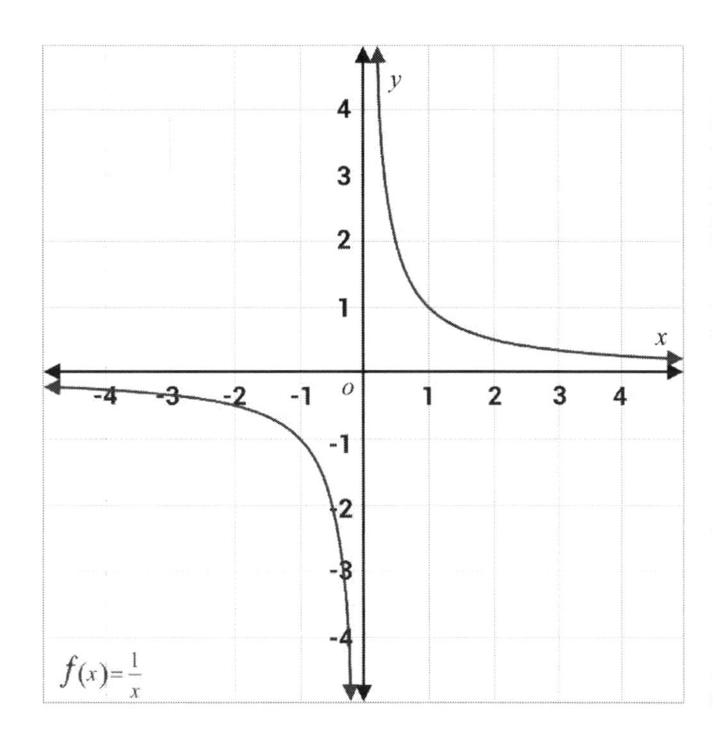

The second graph shows a rational function with a vertical asymptote at x = -.5. Again this can be found by just setting the denominator equal to 0. So, $2x^2 + x = 0, 2x + 1 = 0, 2x = -1, x = -.5$. This graph also has a hole in the graph at $x = 0$. This is because both the numerator and denominator are equal to 0 when $x = 0$.

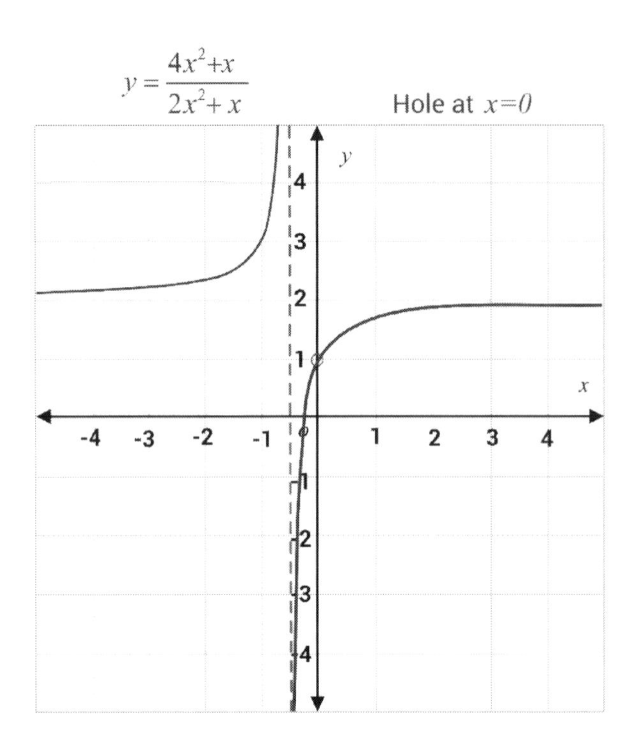

Exponential Functions

An *exponential function* is a function of the form $f(x) = b^x$, where *b* is a positive real number other than 1. In such a function, *b* is called the *base*.

The *domain* of an exponential function is all real numbers, and the *range* is all positive real numbers. There will always be a horizontal asymptote of $y = 0$ on one side. If *b* is greater than 1, then the graph will be increasing moving to the right. If *b* is less than 1, then the graph will be decreasing moving to the right. Exponential functions are one-to-one. The basic exponential function graph will go through the point (0,1).

Example

Solve $5^{x+1} = 25$.

Get the x out of the exponent by rewriting the equation $5^{x+1} = 5^2$ so that both sides have a base of 5.

Since the bases are the same, the exponents must be equal to each other.

This leaves $x + 1 = 2$ or $x = 1$.

To check the answer, the x-value of 1 can be substituted back into the original equation.

Logarithmic Functions

A *logarithmic function* is an inverse for an exponential function. The inverse of the base *b* exponential function is written as $\log_b(x)$, and is called the *base b logarithm*. The domain of a logarithm is all positive real numbers. It has the properties that $\log_b(b^x) = x$. For positive real values of *x*, $b^{\log_b(x)} = x$.

When there is no chance of confusion, the parentheses are sometimes skipped for logarithmic functions: $\log_b(x)$ may be written as $\log_b x$. For the special number *e*, the base *e* logarithm is called the *natural logarithm* and is written as $\ln x$. Logarithms are one-to-one.

When working with logarithmic functions, it is important to remember the following properties. Each one can be derived from the definition of the logarithm as the inverse to an exponential function:

$$\log_b 1 = 0$$

$$\log_b b = 1$$

$$\log_b b^p = p$$

$$\log_b MN = \log_b M + \log_b N$$

$$\log_b \frac{M}{N} = \log_b M - \log_b N$$

$$\log_b M^p = p \log_b M$$

When solving equations involving exponentials and logarithms, the following fact should be used:

If *f* is a one-to-one function, $a = b$ is equivalent to $f(a) = f(b)$.

Using this, together with the fact that logarithms and exponentials are inverses, allows manipulations of the equations to isolate the variable.

<u>Example</u>
Solve $4 = \ln(x - 4)$.

Using the definition of a logarithm, the equation can be changed to $e^4 = e^{\ln(x-4)}$.

The functions on the right side cancel with a result of $e^4 = x - 4$.

This then gives $x = 4 + e^4$.

Trigonometric Functions

Trigonometric functions are built out of two basic functions, the *sine* and *cosine*, written as $\sin\theta$ and $\cos\theta$ respectively. Note that similar to logarithms, it is customary to drop the parentheses as long as the result is not confusing.

The sine and cosine are defined using the *unit circle*. If θ is the angle going counterclockwise around the origin from the *x*-axis, then the point on the unit circle in that direction will have the coordinates $(\cos\theta, \sin\theta)$.

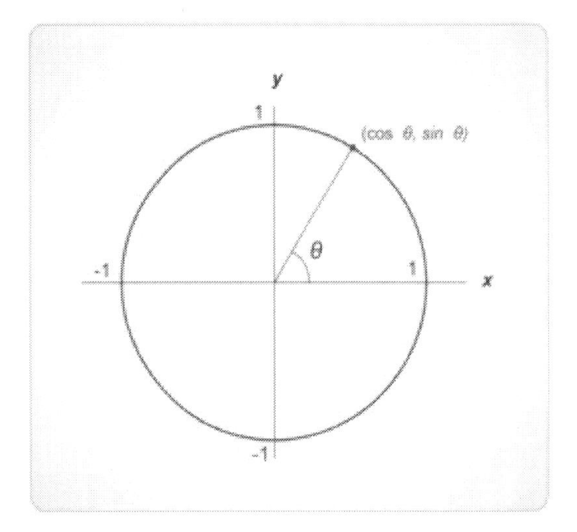

Since the angle returns to the start every 2π radians (or 360 degrees), the graph of these functions will be *periodic*, with period 2π. This means that the graph repeats itself as one moves along the *x*-axis because $\sin\theta = \sin(\theta + 2\pi)$. Cosine is works similarly.

From the unit circle definition, the sine function starts at 0 when $\theta = 0$. It grows to 1 as θ grows to $\pi/2$, and then back to 0 at $\theta = \pi$. Then it decreases to -1 as θ grows to $3\pi/2$, and back up to 0 at $\theta = 2\pi$.

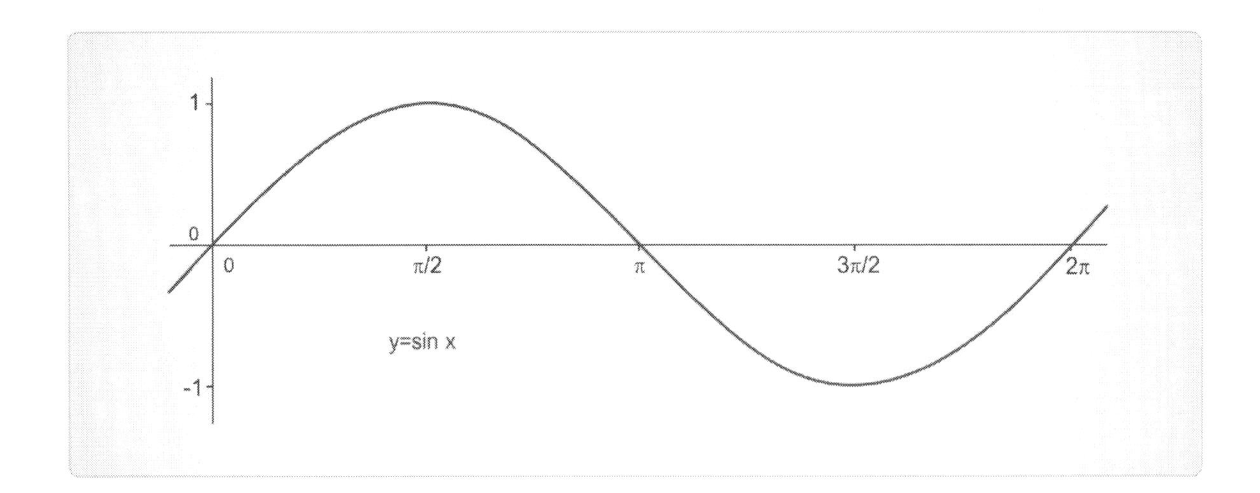

The graph of the cosine is similar. The cosine will start at 1, decreasing to 0 at $\pi/2$ and continuing to decrease to -1 at $\theta = \pi$. Then, it grows to 0 as θ grows to $3\pi/2$ and back up to 1 at $\theta = 2\pi$.

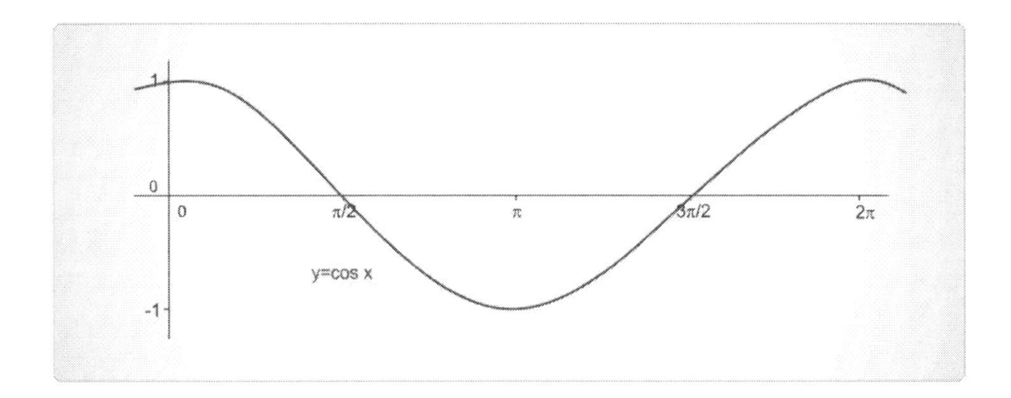

Another trigonometric function, which is frequently used, is the *tangent* function. This is defined as the following equation: $\tan\theta = \frac{\sin\theta}{\cos\theta}$.

The tangent function is a period of π rather than 2π because the sine and cosine functions have the same absolute values after a change in the angle of π, but flip their signs. Since the tangent is a ratio of the two functions, the changes in signs cancel.

The tangent function will be zero when the sine is zero, and it will have a vertical asymptote whenever cosine is zero. The following graph shows the tangent function:

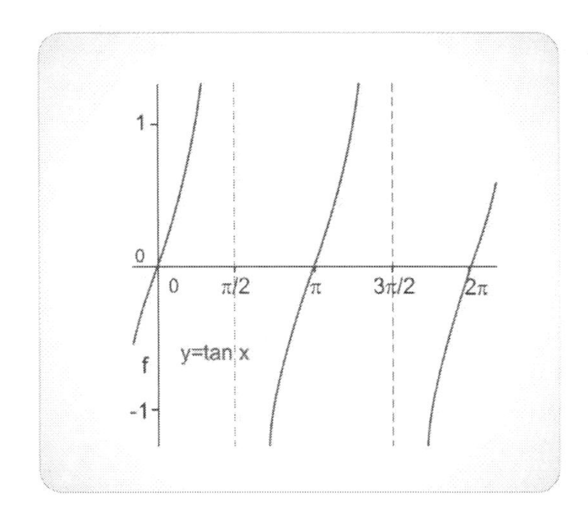

Three other trigonometric functions are sometimes useful. These are the *reciprocal* trigonometric functions, so named because they are just the reciprocals of sine, cosine, and tangent. They are the *cosecant*, defined as $\csc\theta = \frac{1}{\sin\theta}$, the *secant*, $\sec\theta = \frac{1}{\cos\theta}$, and the *cotangent*, $\cot\theta = \frac{1}{\tan\theta}$. Note that from the definition of tangent, $\cot\theta = \frac{\cos\theta}{\sin\theta}$.

In addition, there are three identities that relate the trigonometric functions to one another:

$$\cos\theta = \sin(\frac{\pi}{2} - \theta)$$

$$\csc\theta = \sec\left(\frac{\pi}{2} - \theta\right)$$

$$\cot\theta = \tan(\frac{\pi}{2} - \theta)$$

Here is a list of commonly-needed values for trigonometric functions, given in radians, for the first quadrant:

Table for trigonometric functions

$\sin 0 = 0$	$\cos 0 = 1$	$\tan 0 = 0$
$\sin\frac{\pi}{6} = \frac{1}{2}$	$\cos\frac{\pi}{6} = \frac{\sqrt{3}}{2}$	$\tan\frac{\pi}{6} = \frac{\sqrt{3}}{3}$
$\sin\frac{\pi}{4} = \frac{\sqrt{2}}{2}$	$\cos\frac{\pi}{4} = \frac{\sqrt{2}}{2}$	$\tan\frac{\pi}{4} = 1$
$\sin\frac{\pi}{3} = \frac{\sqrt{3}}{2}$	$\cos\frac{\pi}{3} = \frac{1}{2}$	$\tan\frac{\pi}{3} = \sqrt{3}$
$\sin\frac{\pi}{2} = 1$	$\cos\frac{\pi}{2} = 0$	$\tan\frac{\pi}{2} = undefined$
$\csc 0 = undefined$	$\sec 0 = 1$	$\cot 0 = undefined$
$\csc\frac{\pi}{6} = 2$	$\sec\frac{\pi}{6} = \frac{2\sqrt{3}}{3}$	$\cot\frac{\pi}{6} = \sqrt{3}$
$\csc\frac{\pi}{4} = \sqrt{2}$	$\sec\frac{\pi}{4} = \sqrt{2}$	$\cot\frac{\pi}{4} = 1$
$\csc\frac{\pi}{3} = \frac{2\sqrt{3}}{3}$	$\sec\frac{\pi}{3} = 2$	$\cot\frac{\pi}{3} = \frac{\sqrt{3}}{3}$
$\csc\frac{\pi}{2} = 1$	$\sec\frac{\pi}{2} = undefined$	$\cot\frac{\pi}{2} = 0$

To find the trigonometric values in other quadrants, complementary angles can be used. The *complementary angle* is the smallest angle between the *x*-axis and the given angle.

Once the complementary angle is known, the following rule is used:

For an angle θ with complementary angle x, the absolute value of a trigonometry function evaluated at θ is the same as the absolute value when evaluated at x.

The correct sign is used based on the functions sine and cosine are given by the *x* and *y* coordinates on the unit circle.

Sine will be positive in quadrants I and II and negative in quadrants III and IV.

Cosine will be positive in quadrants I and IV, and negative in II and III.

Tangent will be positive in I and III, and negative in II and IV.

The signs of the reciprocal functions will be the same as the sign of the function of which they are a reciprocal.

Example

Find $\sin\frac{3\pi}{4}$.

First, the complementary angle must be found.

This angle is in the II quadrant, and the angle between it and the *x*-axis is $\frac{\pi}{4}$.

Now, $\sin\frac{\pi}{4} = \frac{\sqrt{2}}{2}$.

Since this is in the II quadrant, sine takes on positive values (the *y* coordinate is positive in the II quadrant).

Therefore, $\sin\frac{3\pi}{4} = \frac{\sqrt{2}}{2}$.

In addition to the six trigonometric functions defined above, there are inverses for these functions. However, since the trigonometric functions are not one-to-one, one can only construct inverses for them on a restricted domain.

Usually, the domain chosen will be $[0, \pi)$ for cosine and $(-\frac{\pi}{2}, \frac{\pi}{2}]$ for sine. The inverse for tangent can use either of these domains. The inverse functions for the trigonometric functions are also called *arc functions*. In addition to being written with a -1 in the exponent to denote that the function is an inverse, they will sometimes be written with an "a" or "arc" in front of the function name, so $\cos^{-1}\theta = \operatorname{acos}\theta = \arccos\theta$.

When solving equations that involve trigonometric functions, there are often multiple solutions. For example, $2\sin\theta = \sqrt{2}$ can be simplified to $\sin\theta = \frac{\sqrt{2}}{2}$. This has solutions $\theta = \frac{\pi}{4}, \frac{3\pi}{4}$, but in addition, because of the periodicity, any integer multiple of 2π can also be added to these solutions to find another solution.

The full set of solutions is $\theta = \frac{\pi}{4} + 2\pi k, \frac{3\pi}{4} + 2\pi k$ for all integer values of *k*. It is very important to remember to find all possible solutions when dealing with equations that involve trigonometric functions.

The name *trigonometric* comes from the fact that these functions play an important role in the geometry of triangles, particularly right triangles.

Consider the right triangle shown in this figure:

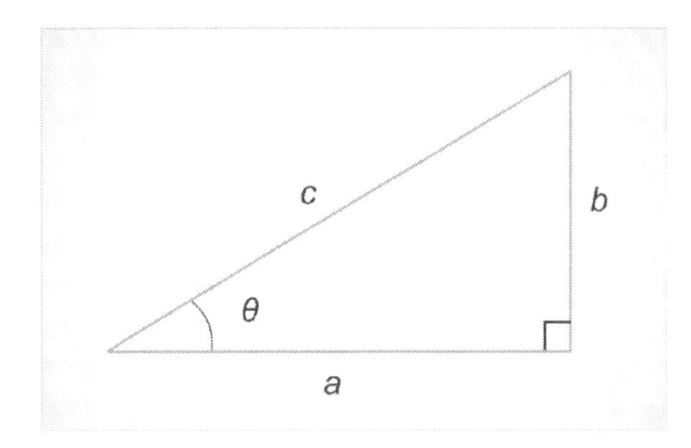

The following hold true:

- $c \sin\theta = b$
- $c \cos\theta = a$
- $\tan\theta = \frac{b}{a}$
- $b \csc\theta = c$
- $a \sec\theta = c$
- $\cot\theta = \frac{a}{b}$

Remember also the angles of a triangle must add up to π radians (180 degrees).

Geometry and Measurement

Plane Geometry

Locations on the plane that have no width or breadth are called *points*. These points usually will be denoted with capital letters such as *P*.

Any pair of points *A*, *B* on the plane will determine a unique straight line between them. This line is denoted *AB*. Sometimes to emphasize a line is being considered, this will be written as $\overleftrightarrow{AB}$.

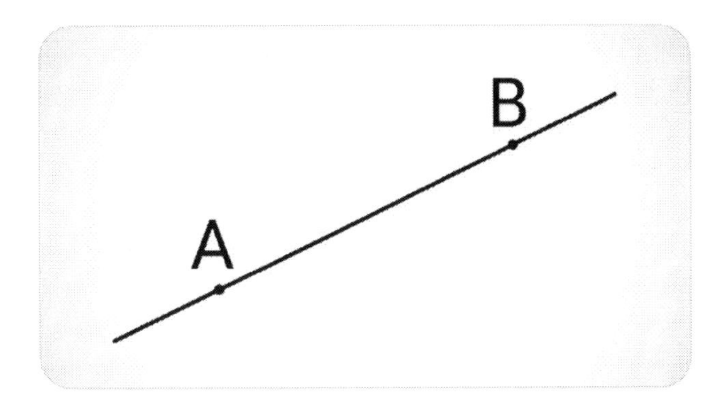

If the Cartesian coordinates for *A* and *B* are known, then the distance $d(A, B)$ along the line between them can be measured using the *Pythagorean formula*, which states that if $A = (x_1, y_1)$ and $B = (x_2, y_2)$, then the distance between them is:

$$d(A, B) = \sqrt{(x_2 - x_1)^2 + (y_2 - y_1)^2}$$

The part of a line that lies between *A* and *B* is called a *line segment*. It has two endpoints, one at *A* and one at *B*. *Rays* also can be formed. Given points *A* and *B*, a *ray* is the portion of a line that starts at one of these points, passes through the other, and keeps on going. Therefore, a ray has a single endpoint, but the other end goes off to infinity.

Given a pair of points *A* and *B*, a circle centered at *A* and passing through *B* can be formed. This is the set of points whose distance from *A* is exactly $d(A, B)$. The radius of this circle will be $d(A, B)$.

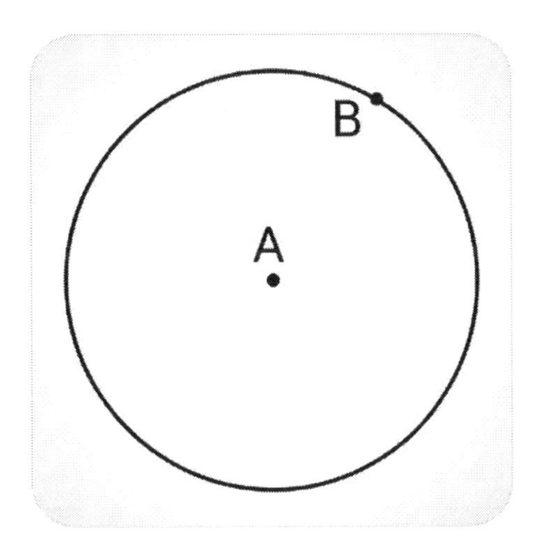

The *circumference* of a circle is the distance traveled by following the edge of the circle for one complete revolution, and the length of the circumference is given by $2\pi r$, where *r* is the radius of the circle. The formula for circumference is $C = 2\pi r$.

When two lines cross, they form an *angle*. The point where the lines cross is called the *vertex* of the angle. The angle can be named by either just using the vertex, $\angle A$, or else by listing three points $\angle BAC$, as shown in the diagram below.

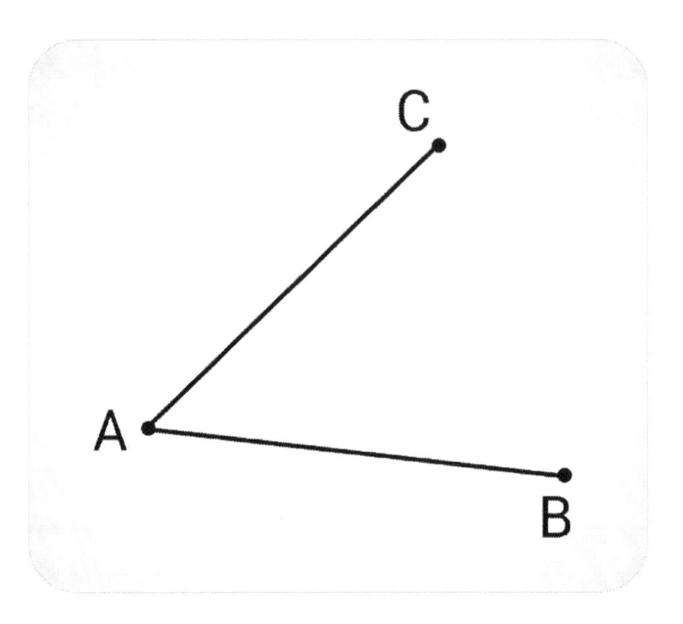

The measurement of an angle can be given in degrees or in radians. In degrees, a full circle is 360 degrees, written 360°. In radians, a full circle is 2π radians.

Given two points on the circumference of a circle, the path along the circle between those points is called an *arc* of the circle. For example, the arc between *B* and *C* is denoted by a thinner line:

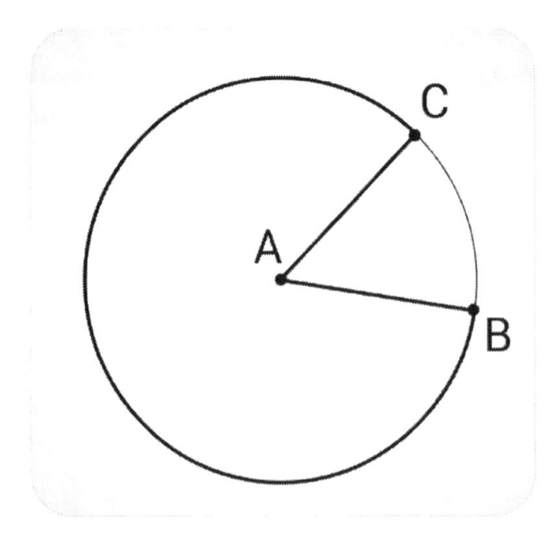

The length of the path along an arc is called the *arc length*. If the circle has radius *r*, then the arc length is given by multiplying the measure of the angle in radians by the radius of the circle.

Two lines are said to be *parallel* if they never intersect. If the lines are *AB* and *CD*, then this is written as $AB \parallel CD$.

If two lines cross to form four quarter-circles, that is, 90° angles, the two lines are *perpendicular*. If the point at which they cross is *B*, and the two lines are *AB* and *BC*, then this is written as $AB \perp BC$.

A *polygon* is a closed figure (meaning it divides the plane into an inside and an outside) consisting of a collection of line segments between points. These points are called the *vertices* of the polygon. These line segments must not overlap one another. Note that the number of sides is equal to the number of angles, or vertices of the polygon. The angles between line segments meeting one another in the polygon are called *interior angles*.

A *regular polygon* is a polygon whose edges are all the same length and whose interior angles are all of equal measure.

A *triangle* is a polygon with three sides. A *quadrilateral* is a polygon with four sides.

A *right triangle* is a triangle that has one 90° angle.

The sum of the interior angles of any triangle must add up to 180°.

An *isosceles triangle* is a triangle in which two of the sides are the same length. In this case, it will always have two congruent interior angles. If a triangle has two congruent interior angles, it will always be isosceles.

An *equilateral triangle* is a triangle whose sides are all the same length and whose angles are all equivalent to one another, equal to 60°. Equilateral triangles are examples of regular polygons. Note that equilateral triangles are also isosceles.

A *rectangle* is a quadrilateral whose interior angles are all 90°. A rectangle has two sets of sides that are equal to one another.

A *square* is a rectangle whose width and height are equal. Therefore, squares are regular polygons.

A *parallelogram* is a quadrilateral in which the opposite sides are parallel and equivalent to each other.

The Coordinate Plane

The coordinate plane can be divided into four *quadrants*. The upper-right part of the plane is called the *first quadrant*, where both *x* and *y* are positive. The *second quadrant* is the upper-left, where *x* is negative but *y* is positive. The *third quadrant* is the lower left, where both *x* and *y* are negative. Finally, the *fourth quadrant* is in the lower right, where *x* is positive but *y* is negative. These quadrants are often written with Roman numerals:

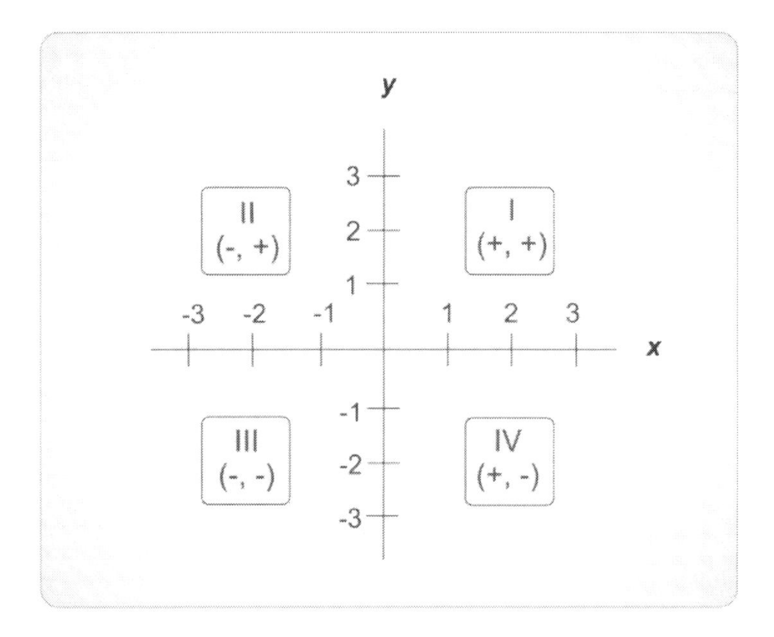

In addition to graphing individual points as shown above, the graph lines and curves in the plane can be graphed corresponding to equations. In general, if there is some equation involving *x* and *y*, then the *graph* of that equation consists of all the points (x, y) in the Cartesian coordinate plane, which satisfy this equation.

Given the equation $y = x + 2$, the point $(0, 2)$ is in the graph, since $2 = 0 + 2$ is a true equation. However, the point $(1, 4)$ will *not* be in the graph, because $4 = 1 + 2$ is false.

Straight Lines

The simplest equations to graph are the equations whose graphs are lines, called *linear equations*. Every linear equation can be rewritten algebraically so that it looks like $Ax + By = C$.

First, the ratio of the change in the *y* coordinate to the change in the *x* coordinate is constant for any two distinct points on the line. In any pair of points on a line, two points, (x_1, y_1) and (x_2, y_2)—

where $x_1 \neq x_2$—the ratio $\frac{y_2-y_1}{x_2-x_1}$ will always be the same, even if another pair of points is used.

This ratio, $\frac{y_2-y_1}{x_2-x_1}$, is called the *slope* of the line and is often denoted with the letter m. If the slope is *positive*, then the line goes upward when moving to the right. If the slope is *negative*, then it moves downward when moving to the right. If the slope is 0, then the line is *horizontal*, and the *y* coordinate is constant along the entire line. For lines where the *x* coordinate is constant along the entire line, the slope is not defined, and these lines are called *vertical* lines.

The *y* coordinate of the point where the line touches the *y*-axis is called the *y-intercept* of the line. It is often denoted by the letter b, used in the form of the linear equation $y = mx + b$. The *x* coordinate of the point where the line touches the *x*-axis is called the *x-intercept*. It is also called the *zero* of the line.

Suppose two lines have slopes m_1 and m_2. If the slopes are equal, $m_1 = m_2$, then the lines are *parallel*. Parallel lines never meet one another. If $m_1 = -\frac{1}{m_2}$, then the lines are called *perpendicular* or *orthogonal*. Their slopes can also be called opposite reciprocals of each other.

There are several convenient ways to write down linear equations. The common forms are listed here:

Standard Form: $Ax + By = C$, where the slope is given by $\frac{-A}{B}$, and the *y*-intercept is given by $\frac{C}{B}$.

Slope-Intercept Form: $y = mx + b$, where the slope is *m*, and the *y*-intercept is *b*.

Point-Slope Form: $y - y_1 = m(x - x_1)$, where *m* is the slope, and (x_1, y_1) is any point on the line.

Two-Point Form: $\frac{y-y_1}{x-x_1} = \frac{y_2-y_1}{x_2-x_1}$, where (x_1, y_1), and (x_2, y_2) are any two distinct points on the line.

Intercept Form: $\frac{x}{x_1} + \frac{y}{y_1} = 1$, where x_1 is the *x*-intercept, and y_1 is the *y*-intercept.

Depending upon the given information, different forms of the linear equation can be easier to write down than others. When given two points, the two-point form is easy to write down. If the slope and a single point is known, the point-slope form is easiest to start with. In general, which form to start with depends upon the given information.

Conics

The graph of an equation of the form $y = ax^2 + bx + c$ or $x = ay^2 + by + c$ is called a *parabola*.

The graph of an equation of the form $\frac{x^2}{a^2} - \frac{y^2}{b^2} = 1$ or $-\frac{x^2}{a^2} + \frac{y^2}{b^2} = 1$ is called a *hyperbola*.

The graph of an equation of the form $\frac{(x-x_0)^2}{a^2} + \frac{(y-y_0)^2}{b^2} = 1$ is called an *ellipse*. If $a = b$ then this is a circle with *radius* $r = \frac{1}{a}$.

Sets of Points in the Plane

The *midpoint* between two points, (x_1, y_1) and (x_2, y_2), is given by taking the average of the *x* coordinates and the average of the *y* coordinates: $\left(\frac{x_1+x_2}{2}, \frac{y_1+y_2}{2}\right)$.

The *distance* between two points, (x_1, y_1) and (x_2, y_2), is given by the *Pythagorean formula*, $\sqrt{(x_2 - x_1)^2 + (y_2 - y_1)^2}$.

To find the perpendicular distance between a line $Ax + By = C$ and a point (x_1, y_1) not on the line, we need to use the formula:

$$\frac{|Ax_1 + By_1 + C|}{\sqrt{A^2 + B^2}}$$

Transformations of a Plane

Given a figure drawn on a plane, many changes can be made to that figure, including *rotation*, *translation*, and *reflection*. Rotations turn the figure about a point, translations slide the figure, and reflections flip the figure over a specified line. When performing these transformations, the original figure is called the *pre-image*, and the figure after transformation is called the *image*.

More specifically, *translation* means that all points in the figure are moved in the same direction by the same distance. In other words, the figure is slid in some fixed direction. Of course, while the entire figure is slid by the same distance, this does not change any of the measurements of the figures involved. The result will have the same distances and angles as the original figure.

In terms of Cartesian coordinates, a translation means a shift of each of the original points (x, y) by a fixed amount in the *x* and *y* directions, to become $(x + a, y + b)$.

Another procedure that can be performed is called *reflection*. To do this, a line in the plane is specified, called the *line of reflection*. Then, take each point and flip it over the line so that it is the same distance from the line but on the opposite side of it. This does not change any of the distances or angles involved, but it does reverse the order in which everything appears.

To reflect something over the *x*-axis, the points (x, y) are sent to $(x, -y)$. To reflect something over the *y*-axis, the points (x, y) are sent to the points $(-x, y)$. Flipping over other lines is not something easy to express in Cartesian coordinates. However, by drawing the figure and the line of reflection, the distance to the line and the original points can be used to find the reflected figure.

Example: Reflect this triangle with vertices (-1, 0), (2, 1), and (2, 0) over the *y*-axis. The pre-image is shown below.

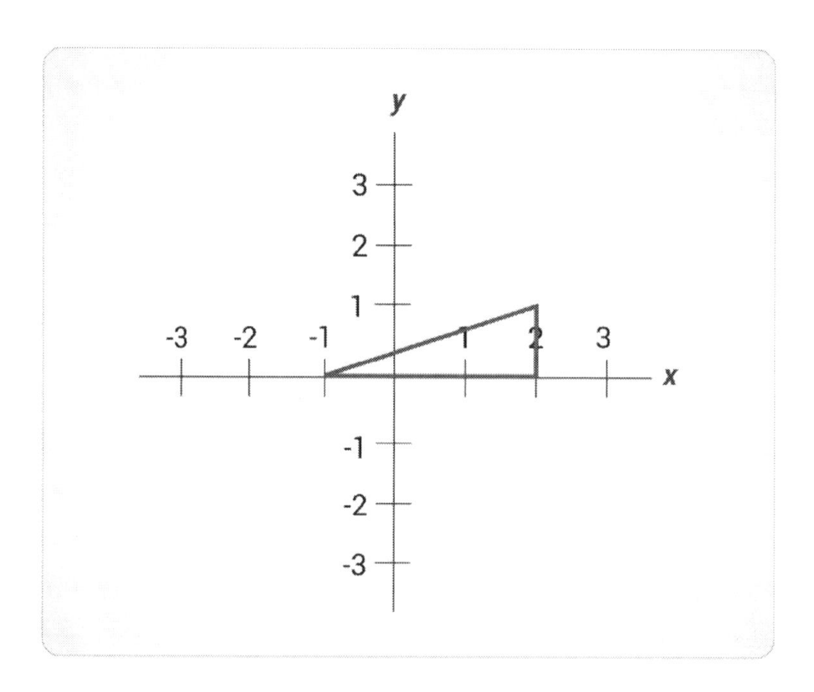

To do this, flip the *x* values of the points involved to the negatives of themselves, while keeping the *y* values the same. The image is shown here.

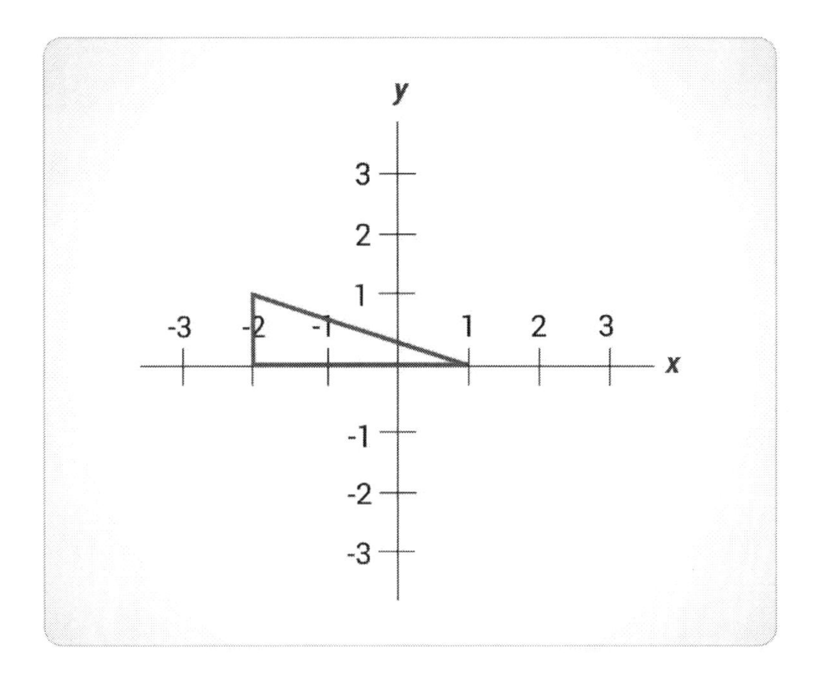

The new vertices will be (1, 0), (-2, 1), and (-2, 0).

Another procedure that does not change the distances and angles in a figure is *rotation*. In this procedure, pick a center point, then rotate every vertex along a circle around that point by the same angle. This procedure is also not easy to express in Cartesian coordinates, and this is not a requirement on this test. However, as with reflections, it's helpful to draw the figures and see what the result of the rotation would look like. This transformation can be performed using a compass and protractor.

Each one of these transformations can be performed on the coordinate plane without changes to the original dimensions or angles.

If two figures in the plane involve the same distances and angles, they are called *congruent figures*. In other words, two figures are congruent when they go from one form to another through reflection, rotation, and translation, or a combination of these.

Remember that rotation and translation will give back a new figure that is identical to the original figure, but reflection will give back a mirror image of it.

To recognize that a figure has undergone a rotation, check to see that the figure has not been changed into a mirror image, but that its orientation has changed (that is, whether the parts of the figure now form different angles with the *x* and *y* axes).

To recognize that a figure has undergone a translation, check to see that the figure has not been changed into a mirror image, and that the orientation remains the same.

To recognize that a figure has undergone a reflection, check to see that the new figure is a mirror image of the old figure.

Keep in mind that sometimes a combination of translations, reflections, and rotations may be performed on a figure.

Dilation

A *dilation* is a transformation that preserves angles, but not distances. This can be thought of as stretching or shrinking a figure. If a dilation makes figures larger, it is called an *enlargement*. If a dilation makes figures smaller, it is called a *reduction*. The easiest example is to dilate around the origin. In this case, multiply the *x* and *y* coordinates by a *scale factor*, k, sending points (x, y) to (kx, ky).

As an example, draw a dilation of the following triangle, whose vertices will be the points (-1, 0), (1, 0), and (1, 1).

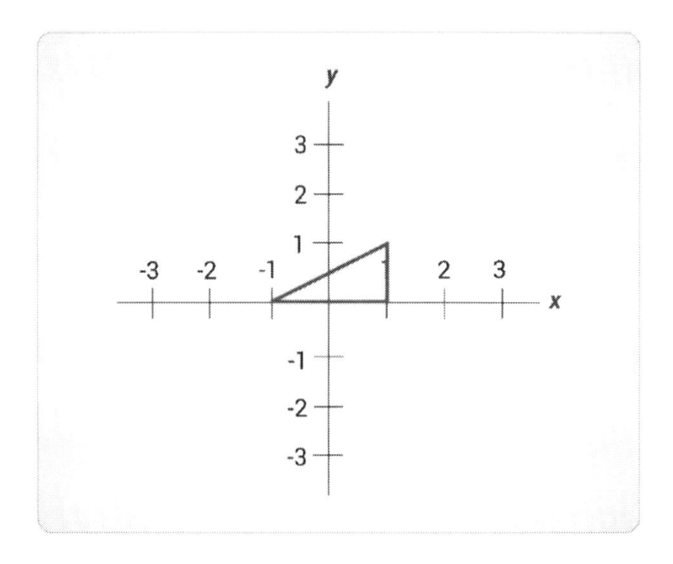

For this problem, dilate by a scale factor of 2, so the new vertices will be (-2, 0), (2, 0), and (2, 2).

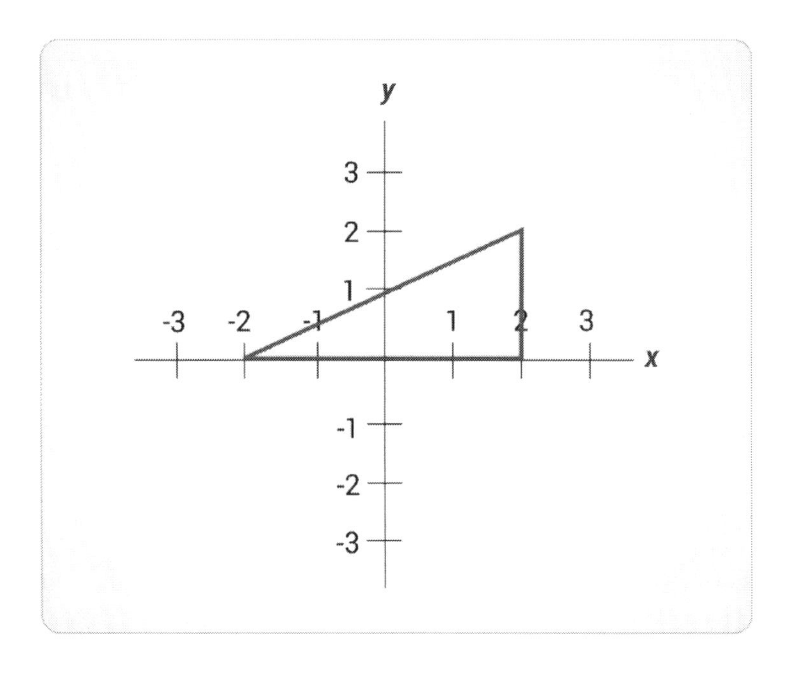

Note that after a dilation, the distances between the vertices of the figure will have changed, but the angles remain the same. The two figures that are obtained by dilation, along with possibly translation, rotation, and reflection, are all *similar* to one another. Another way to think of this is that similar figures have the same number of vertices and edges, and their angles are all the same. Similar figures have the same basic shape, but are different in size.

Symmetry

Using the types of transformations above, if an object can undergo these changes and not appear to have changed, then the figure is symmetrical. If an object can be split in half by a line and flipped over that line to lie directly on top of itself, it is said to have *line symmetry*. An example of both types of figures is seen below.

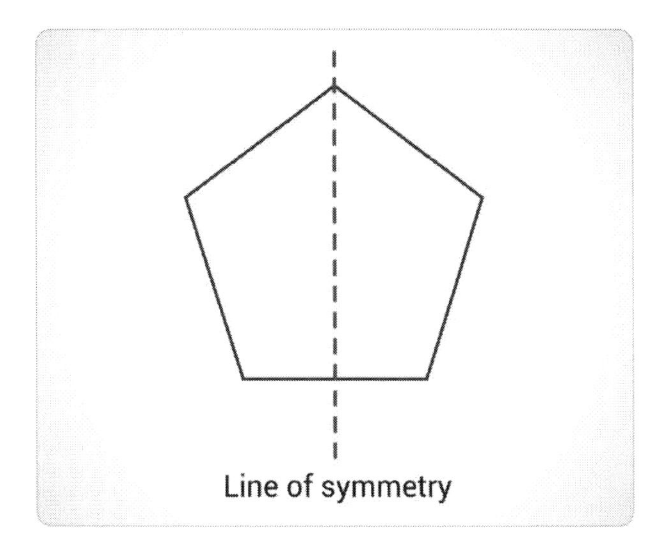

If an object can be rotated about its center to any degree smaller than 360, and it lies directly on top of itself, the object is said to have *rotational symmetry*. An example of this type of symmetry is shown below. The pentagon has an order of 5.

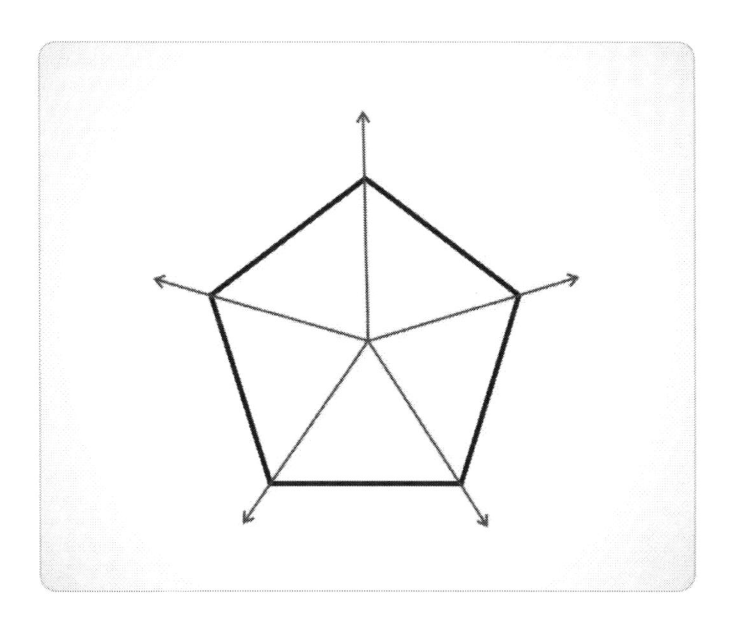

The rotational symmetry lines in the figure above can be used to find the angles formed at the center of the pentagon. Knowing that all of the angles together form a full circle, at 360 degrees, the figure can be split into 5 angles equally. By dividing the 360° by 5, each angle is 72°.

Given the length of one side of the figure, the perimeter of the pentagon can also be found using rotational symmetry. If one side length was 3 cm, that side length can be rotated onto each other side length four times. This would give a total of 5 side lengths equal to 3 cm. To find the perimeter, or distance around the figure, multiply 3 by 5. The perimeter of the figure would be 15 cm.

If a line cannot be drawn anywhere on the object to flip the figure onto itself or rotated less than or equal to 180 degrees to lay on top of itself, the object is asymmetrical. Examples of these types of figures are shown below.

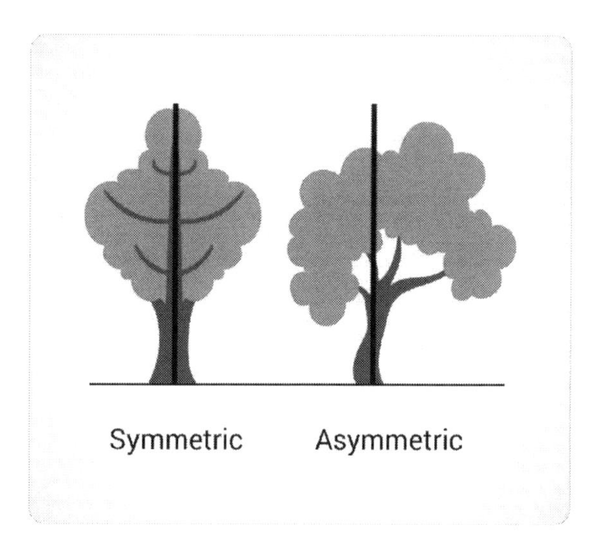

Perimeters and Areas

The *perimeter* of a polygon is the total length of a trip around the whole polygon, starting and ending at the same point. It is found by adding up the lengths of each line segment in the polygon. For a rectangle with sides of length *x* and *y*, the perimeter will be $2x + 2y$.

The area of a polygon is the area of the region that it encloses. Regarding the area of a rectangle with sides of length x and y, the area is given by xy. For a triangle with a base of length b and a height of length h, the area is $\frac{1}{2}bh$.

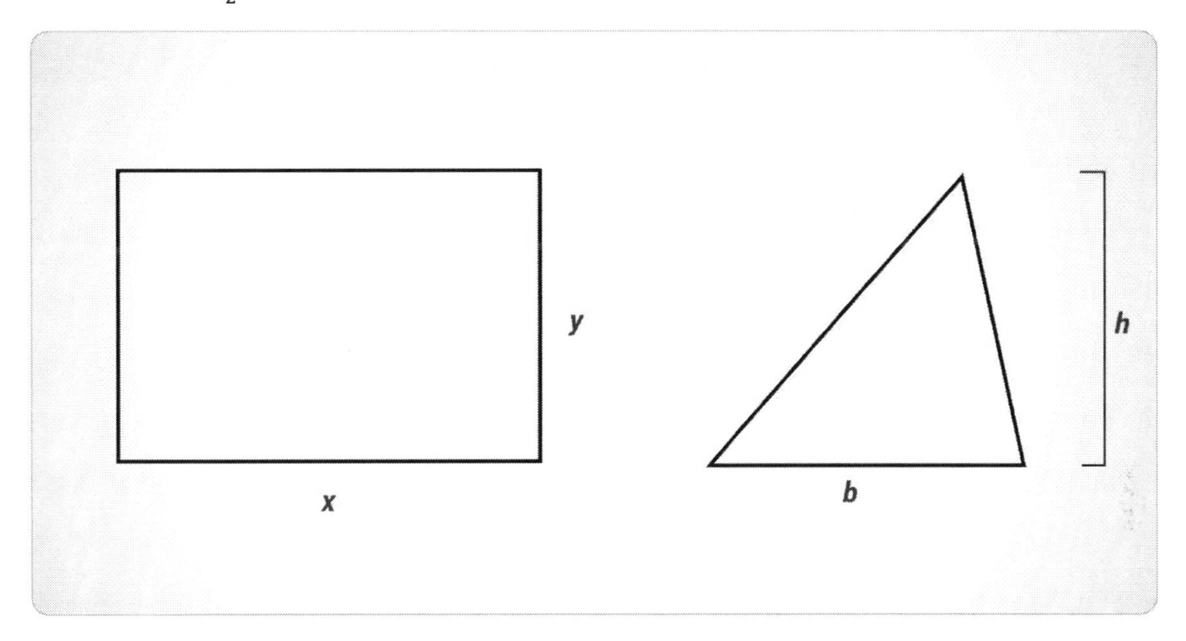

To find the areas of more general polygons, it is usually easiest to break up the polygon into rectangles and triangles. For example, find the area of the following figure whose vertices are (-1, 0), (-1, 2), (1, 3), and (1, 0).

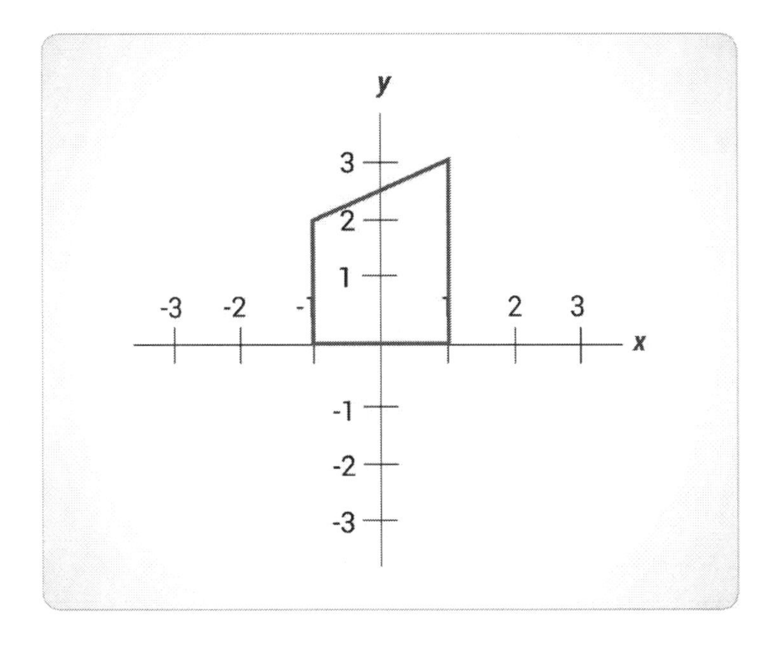

Separate this into a rectangle and a triangle as shown:

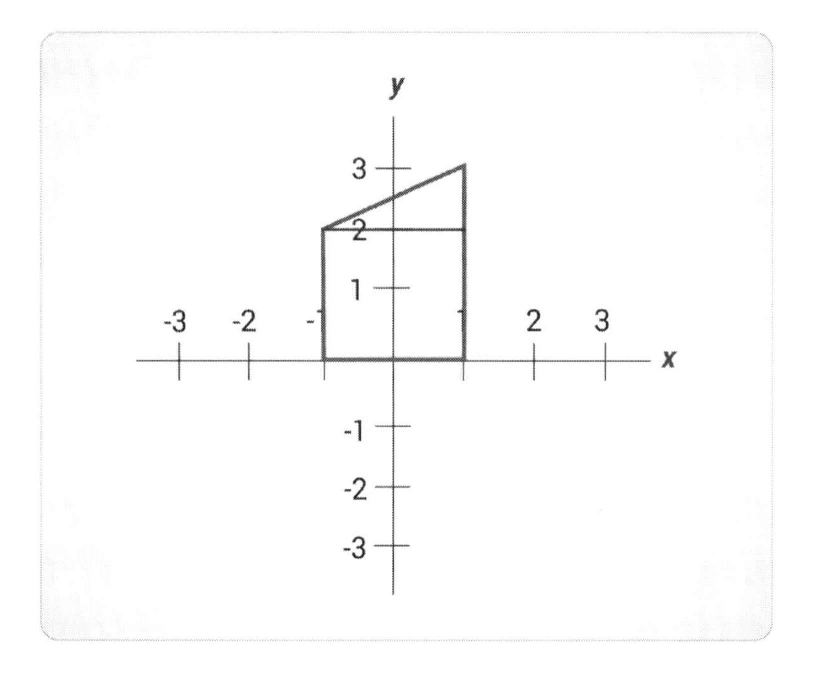

The rectangle has a height of 2 and a width of 2, so it has a total area of $2 \times 2 = 4$. The triangle has a width of 2 and a height of 1, so it has an area of $\frac{1}{2} 2 \times 1 = 1$. Therefore, the entire quadrilateral has an area of 4 + 1 = 5.

As another example, suppose someone wants to tile a rectangular room that is 10 feet by 6 feet using triangular tiles that are 12 inches by 6 inches. How many tiles would be needed? To figure this, first find the area of the room, which will be $10 \times 6 = 60$ square feet. The dimensions of the triangle are 1 foot by ½ foot, so the area of each triangle is $\frac{1}{2} \times 1 \times \frac{1}{2} = \frac{1}{4}$ square feet. Notice that the dimensions of the triangle had to be converted to the same units as the rectangle. Now, take the total area divided by the area of one tile to find the answer: $\frac{60}{\frac{1}{4}} = 60 \times 4 = 240$ tiles required.

Volumes and Surface Areas

Geometry in three dimensions is similar to geometry in two dimensions. The main new feature is that three points now define a unique *plane* that passes through each of them. Three dimensional objects can be made by putting together two dimensional figures in different surfaces. Below, some of the possible three dimensional figures will be provided, along with formulas for their volumes and surface areas.

A rectangular prism is a box whose sides are all rectangles meeting at 90° angles. Such a box has three dimensions: length, width, and height. If the length is *x*, the width is *y*, and the height is *z*, then the volume is given by $V = xyz$.

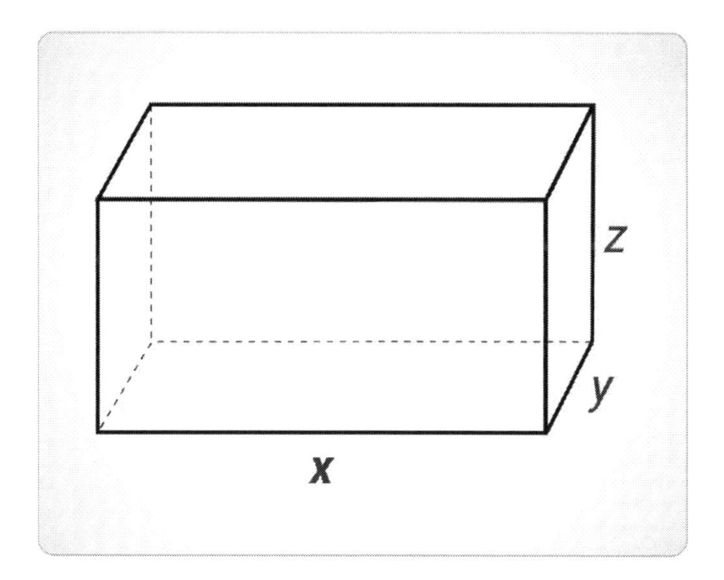

The surface area will be given by computing the surface area of each rectangle and adding them together. There are a total of six rectangles. Two of them have sides of length *x* and *y*, two have sides of length *y* and *z*, and two have sides of length *x* and *z*. Therefore, the total surface area will be given by $SA = 2xy + 2yz + 2xz$.

A *rectangular pyramid* is a figure with a rectangular base and four triangular sides that meet at a single vertex. If the rectangle has sides of length *x* and *y*, then the volume will be given by $V = \frac{1}{3}xyh$.

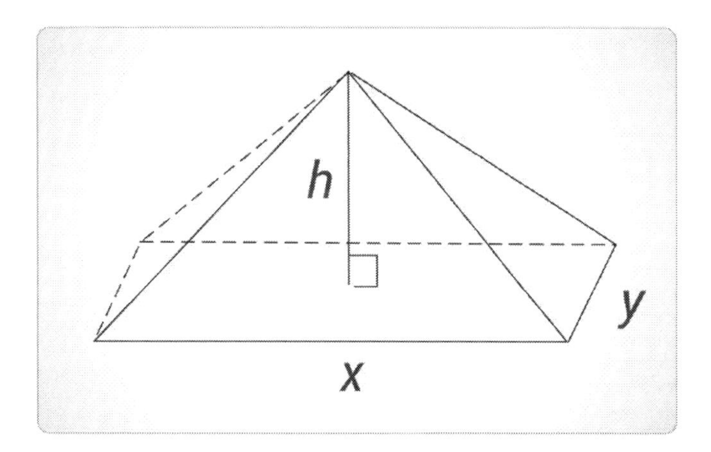

To find the surface area, the dimensions of each triangle need to be known. However, these dimensions can differ depending on the problem in question. Therefore, there is no general formula for calculating total surface area.

A *sphere* is a set of points all of which are equidistant from some central point. It is like a circle, but in three dimensions. The volume of a sphere of radius *r* is given by $V = \frac{4}{3}\pi r^3$. The surface area is given by $A = 4\pi r^2$.

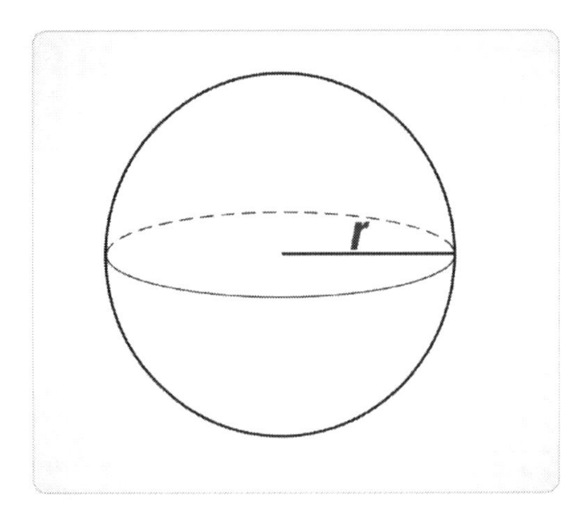

The Pythagorean Theorem

The Pythagorean theorem is an important result in geometry. It states that for right triangles, the sum of the squares of the two shorter sides will be equal to the square of the longest side (also called the *hypotenuse*). The longest side will always be the side opposite to the 90° angle. If this side is called *c*, and the other two sides are *a* and *b*, then the Pythagorean theorem states that $c^2 = a^2 + b^2$. Since lengths are always positive, this also can be written as $c = \sqrt{a^2 + b^2}$.

A diagram to show the parts of a triangle using the Pythagorean theorem is below.

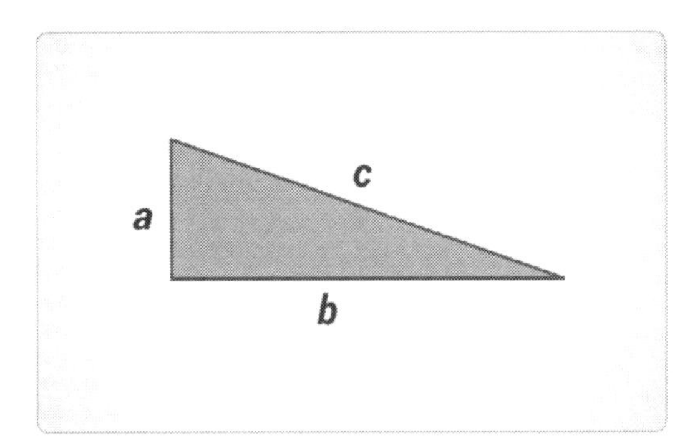

As an example of the theorem, suppose that Shirley has a rectangular field that is 5 feet wide and 12 feet long, and she wants to split it in half using a fence that goes from one corner to the opposite corner. How long will this fence need to be? To figure this out, note that this makes the field into two right triangles, whose hypotenuse will be the fence dividing it in half. Therefore, the fence length will be given by $\sqrt{5^2 + 12^2} = \sqrt{169} = 13$ feet long.

Similar Figures and Proportions

Sometimes, two figures are similar, meaning they have the same basic shape and the same interior angles, but they have different dimensions. If the ratio of two corresponding sides is known, then that ratio, or scale factor, holds true for all of the dimensions of the new figure.

Here is an example of applying this principle. Suppose that Lara is 5 feet tall and is standing 30 feet from the base of a light pole, and her shadow is 6 feet long. How high is the light on the pole? To figure this, it helps to make a sketch of the situation:

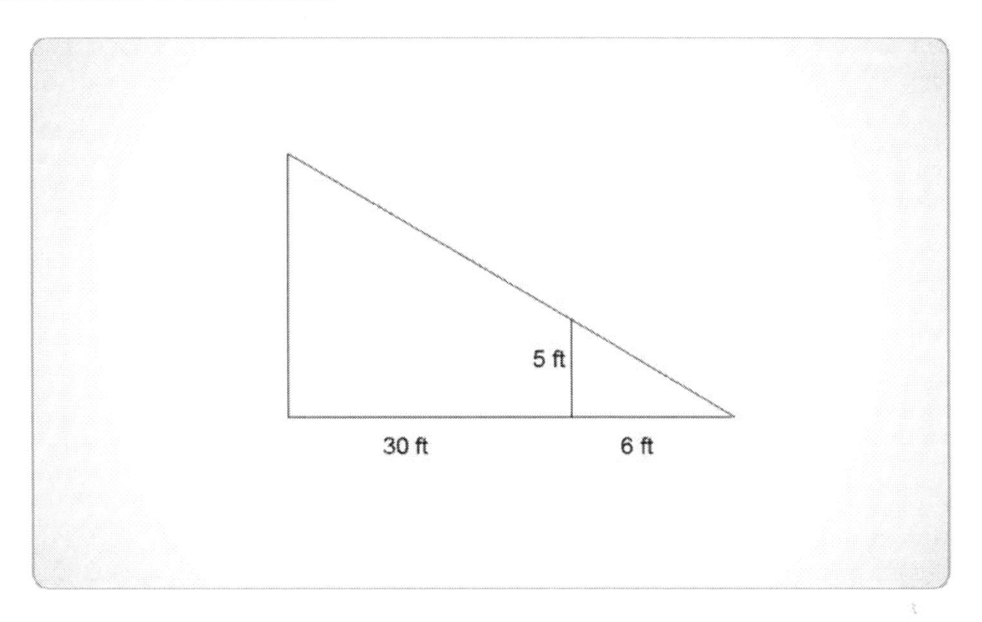

The light pole is the left side of the triangle. Lara is the 5-foot vertical line. Notice that there are two right triangles here, and that they have all the same angles as one another. Therefore, they form similar triangles. So, figure the ratio of proportionality between them.

The bases of these triangles are known. The small triangle, formed by Lara and her shadow, has a base of 6 feet. The large triangle, formed by the light pole along with the line from the base of the pole out to the end of Lara's shadow is $30 + 6 = 36$ feet long. So, the ratio of the big triangle to the little triangle will be $\frac{36}{6} = 6$. The height of the little triangle is 5 feet. Therefore, the height of the big triangle will be $6 \times 5 = 30$ feet, meaning that the light is 30 feet up the pole.

Notice that the perimeter of a figure changes by the ratio of proportionality between two similar figures, but the area changes by the *square* of the ratio. This is because if the length of one side is doubled, the area is quadrupled.

As an example, suppose two rectangles are similar, but the edges of the second rectangle are three times longer than the edges of the first rectangle. The area of the first rectangle is 10 square inches. How much more area does the second rectangle have than the first?

To answer this, note that the area of the second rectangle is $3^2 = 9$ times the area of the first rectangle, which is 10 square inches. Therefore, the area of the second rectangle is going to be $9 \times 10 = 90$ square inches. This means it has $90 - 10 = 80$ square inches more area than the first rectangle.

As a second example, suppose *X* and *Y* are similar right triangles. The hypotenuse of *X* is 4 inches. The area of *Y* is $\frac{1}{4}$ the area of *X*. What is the hypotenuse of *Y*?

First, realize the area has changed by a factor of $\frac{1}{4}$. The area changes by a factor that is the *square* of the ratio of changes in lengths, so the ratio of the lengths is the square root of the ratio of areas. That means that the ratio of lengths must be is $\sqrt{\frac{1}{4}} = \frac{1}{2}$, and the hypotenuse of *Y* must be $\frac{1}{2} \times 4 = 2$ inches.

Volumes between similar solids change like the cube of the change in the lengths of their edges. Likewise, if the ratio of the volumes between similar solids is known, the ratio between their lengths is known by finding the cube root of the ratio of their volumes.

For example, suppose there are two similar rectangular pyramids *X* and *Y*. The base of *X* is 1 inch by 2 inches, and the volume of *X* is 8 inches. The volume of *Y* is 64 inches. What are the dimensions of the base of *Y*?

To answer this, first find the ratio of the volume of *Y* to the volume of *X*. This will be given by $\frac{64}{8} = 8$. Now the ratio of lengths is the cube root of the ratio of volumes, or $\sqrt[3]{8} = 2$. So, the dimensions of the base of *Y* must be 2 inches by 4 inches.

Geometric Reasoning and Graphing

The four-step process of problem solving can be used with geometric reasoning problems as well. There are many geometric properties and terminology included within geometric reasoning.

For example, the perimeter of a rectangle can be written in the terms of the width, or the width can be written in terms of the length.

Example

The width of a rectangle is 2 centimeters less than the length. If the perimeter of the rectangle is 44 centimeters, then what are the dimensions of the rectangle?

The first step is to determine the unknown, which is in terms of the length, l.

The second step is to translate the problem into the equation using the perimeter of a rectangle, $P = 2l + 2w$. The width is the length minus 2 centimeters. The resulting equation is $2l + 2(l - 2) = 44$. The equation can be solved as follows:

$2l + 2l - 4 = 44$	Apply the distributive property on the left side of the equation
$4l - 4 = 44$	Combine like terms on the left side of the equation
$4l = 48$	Add 4 to both sides of the equation
$l = 12$	Divide both sides of the equation by 4

The length of the rectangle is 12 centimeters. The width is the length minus 2 centimeters, which is 10 centimeters. Checking the answers for length and width forms the following equation: $44 = 2(12) + 2(10)$. The equation can be solved using the order of operations to form a true statement: $44 = 44$.

Equations can also be created from complementary angles (angles that add up to 90°) and supplementary angles (angles that add up to 180°).

Example

Two angles are complementary. If one angle is four times the other angle, what is the measure of each angle?

The first step is to determine the unknown, which is the measure of the angle.

The second step is to translate the problem into the equation using the known statement: the sum of two complementary angles is 90°. The resulting equation is $4x + x = 90$. The equation can be solved as follows:

$5x = 90$	Combine like terms on the left side of the equation
$x = 18$	Divide both sides of the equation by 5

The first angle is 18° and the second angle is 4 times the unknown, which is 4 times 18 or 72°.

Going back to check the answer with the original question, 72 and 18 have a sum of 90, making them complementary angles. Seventy-two degrees is also four times the other angle, 18 degrees.

Data Analysis, Statistics, Probability, and College-Level Concepts

What are Statistics?

The field of statistics describes relationships between quantities that are related, but not necessarily in a deterministic manner. For example, a graduating student's salary will often be higher when the student graduates with a higher GPA, but this is not always the case. Likewise, people who smoke tobacco are more likely to develop lung cancer, but, in fact, it is possible for non-smokers to develop the disease as well. *Statistics* describes these kinds of situations, where the likelihood of some outcome depends on the starting data.

Descriptive statistics involves analyzing a collection of data to describe its broad properties such average (or mean), what percent of the data falls within a given range, and other such properties. An example of this would be taking all of the test scores from a given class and calculating the average test score. *Inferential statistics* attempts to use data about a subset of some population to make inferences about the rest of the population. An example of this would be taking a collection of students who received tutoring and comparing their results to a collection of students who did not receive tutoring, then using that comparison to try to predict whether the tutoring program in question is beneficial.

To be sure that inferences have a high probability of being true for the whole population, the subset that is analyzed needs to resemble a miniature version of the population as closely as possible. For this reason, statisticians like to choose random samples from the population to study, rather than picking a specific group of people based on some similarity. For example, studying the incomes of people who live in Portland does not tell anything useful about the incomes of people who live in Tallahassee.

Mean, Median, and Mode

Mean

Suppose that you have a set of data points and some description of the general properties of this data need to be found.

The first property that can be defined for this set of data is the *mean*. This is the same as average. To find the mean, add up all the data points, then divide by the total number of data points. For example, suppose that in a class of 10 students, the scores on a test were 50, 60, 65, 65, 75, 80, 85, 85, 90, 100. Therefore, the average test score will be:

$$\frac{50 + 60 + 65 + 65 + 75 + 80 + 85 + 85 + 90 + 100}{10} = 75.5$$

The mean is a useful number if the distribution of data is normal (more on this later), which roughly means that the frequency of different outcomes has a single peak and is roughly equally distributed on both sides of that peak. However, it is less useful in some cases where the data might be split or where there are some *outliers*. Outliers are data points that are far from the rest of the data. For example, suppose there are 10 executives and 90 employees at a company. The executives make $1000 per hour, and the employees make $10 per hour.

Therefore, the average pay rate will be:

$$\frac{\$1000 \times 10 + \$10 \times 90}{100} = \$109 \text{ per hour}$$

In this case, this average is not very descriptive since it's not close to the actual pay of the executives or the employees.

Median

Another useful measurement is the *median*. In a data set, the median is the point in the middle. The middle refers to the point where half the data comes before it and half comes after, when the data is recorded in numerical order. For instance, these are the speeds of the fastball of a pitcher during the last inning that he pitched (in order from least to greatest):

90, 92, 93, 93, 95, 96, 97, 97, 97

There are nine total numbers, so the middle or *median* number is the 5th one, which is 95.

In cases where the number of data points is an even number, then the average of the two middle points is taken. In the previous example of test scores, the two middle points are 75 and 80. Since there is no single point, the average of these two scores needs to be found. The average is:

$$\frac{75 + 80}{2} = 77.5$$

The median is generally a good value to use if there are a few outliers in the data. It prevents those outliers from affecting the "middle" value as much as when using the mean.

Since an outlier is a data point that is far from most of the other data points in a data set, this means an outlier also is any point that is far from the median of the data set. The outliers can have a substantial effect on the mean of a data set, but they usually do not change the median or mode, or do not change them by a large quantity. For example, consider the data set (3, 5, 6, 6, 6, 8). This has a median of 6 and a mode of 6, with a mean of $\frac{34}{6} \approx 5.67$. Now, suppose a new data point of 1000 is added so that the data set is now (3, 5, 6, 6, 6, 8, 1000). This does not change the median or mode, which are both still 6.

However, the average is now $\frac{1034}{7}$, which is approximately 147.7. In this case, the median and mode will be better descriptions for most of the data points.

The reason for outliers in a given data set is a complicated problem. It is sometimes the result of an error by the experimenter, but often they are perfectly valid data points that must be taken into consideration.

Mode

One additional measure to define for *X* is the *mode*. This is the data point that appears most frequently. If two or more data points all tie for the most frequent appearance, then each of them is considered a mode. In the case of the test scores, where the numbers were 50, 60, 65, 65, 75, 80, 85, 85, 90, 100, there are two modes: 65 and 85.

Quartiles and Percentiles

The *first quartile* of a set of data *X* refers to the largest value from the first ¼ of the data points. In practice, there are sometimes slightly different definitions that can be used, such as the median of the first half of the data points (excluding the median itself if there are an odd number of data points). The term also has a slightly different use: when it is said that a data point lies *in the first quartile*, it means it is less than or equal to the median of the first half of the data points. Conversely, if it lies *at* the first quartile, then it is equal to the first quartile.

When it is said that a data point lies in the *second quartile*, it means it is between the first quartile and the median.

The *third quartile* refers to data that lies between ½ and ¾ of the way through the data set. Again, there are various methods for defining this precisely, but the simplest way is to include all of the data that lie between the median and the median of the top half of the data.

Data that lies in the *fourth quartile* refers to all of the data above the third quartile.

Percentiles may be defined in a similar manner to quartiles. Generally, this is defined in the following manner:

If a data point lies *in the n-th percentile*, this means it lies in the range of the first *n*% of the data.

If a data point lies *at* the *n*-th percentile, then it means that *n*% of the data lies below this data point.

Standard Deviation

Given a data set *X* consisting of data points $(x_1, x_2, x_3, \ldots x_n)$, the *variance* of *X* is defined to be:

$$\frac{\sum_{i=1}^{n}(x_i - \overline{X})^2}{n}$$

This means that the variance of X is the average of the squares of the differences between each data point and the mean of X. In the formula, $\overline{X}$ is the mean of the values in the data set, and x_i represents each individual value in the data set. The sigma notation indicates that the sum should be found with n being the number of values to add together. $i = 1$ means that the values should begin with the first value.

Given a data set *X* consisting of data points $(x_1, x_2, x_3, \dots x_n)$, the *standard deviation* of *X* is defined to be:

$$s_x = \sqrt{\frac{\sum_{i=1}^{n}(x_i - \bar{X})^2}{n}}$$

In other words, the standard deviation is the square root of the variance.

Both the variance and the standard deviation are measures of how much the data tend to be spread out. When the standard deviation is low, the data points are mostly clustered around the mean. When the standard deviation is high, this generally indicates that the data are quite spread out, or else that there are a few substantial outliers.

As a simple example, compute the standard deviation for the data set (1, 3, 3, 5). First, compute the mean, which will be $\frac{1+3+3+5}{4} = \frac{12}{4} = 3$. Now, find the variance of *X* with the formula:

$$\sum_{i=1}^{4}(x_i - \bar{X})^2 = (1-3)^2 + (3-3)^2 + (3-3)^2 + (5-3)^2$$

$$-2^2 + 0^2 + 0^2 + 2^2 = 8$$

Therefore, the variance is $\frac{8}{4} = 2$. Taking the square root, the standard deviation will be $\sqrt{2}$.

Note that the standard deviation only depends upon the mean, not upon the median or mode(s). Generally, if there are multiple modes that are far apart from one another, the standard deviation will be high. A high standard deviation does not always mean there are multiple modes, however.

Fitting Functions to Data

Sometimes, when data are measured, it is not simply measuring the frequency of a given outcome, but rather measuring a relationship between two different quantities. In these cases, there is usually one variable that is controlled, the *independent variable*, and one that depends on this variable, the *dependent variable*. If there is a relationship between the two variables, then they are said to be *correlated*.

There are two caveats to these terms. First, the independent variable is not necessarily controlled by the experimenters. It is simply the one chosen to organize the data. In other words, the data are divided up based on the value of an independent variable. Second, finding a significant relationship between the dependent variable and the independent variable does not necessarily imply that there is a causal relationship between the two variables. It only means that once the independent variable is known, a fairly accurate prediction of the dependent variable can be made. This is often expressed with the phrase *correlation does not imply causation*. In other words, just because there is a relationship between two variables does not mean that one is the cause of the other. There could be other factors involved that are the real cause.

Consider some examples. An experimenter could do an experiment in which the independent variable is the number of hours that a student studies for a given test, and the dependent variable is the score the student receives when he or she actually takes the test. Such an experiment would attempt to measure whether there is a relationship between the time spent studying and the score a student receives when taking the test. The expectation would be that the larger value of the independent variable would yield a larger value for the dependent variable. Another experimenter might do an experiment with runners, where the independent variable is the length of the runner's leg, and the dependent variable is the time it takes for the runner to run a fixed distance. In this experiment, as the independent variable increases, the dependent variable would be expected to decrease.

As an example of the phenomenon that correlation does not imply causation, consider an experiment where the independent variable is the value of a person's house, and the dependent variable is their income. Although people in more expensive houses are expected to make more money, it is clear that their expensive houses are not the cause of them making more money. This illustrates one example of why it is important for experimenters to be careful when drawing conclusions about causation from their data.

Linear Data Fitting

The simplest type of correlation between two variables is a *linear correlation*. If the independent variable is *x* and the dependent variable is *y*, then a linear correlation means $y = mx + b$. If *m* is positive, then *y* will increase as *x* increases. While if *m* is negative, then *y* decreases while *x* increases. The variable *b* represents the value of *y* when *x* is 0.

As one example of such a correlation, consider a manufacturing plant. Suppose *x* is the number of units produced by the plant, and *y* is the cost to the company. In this example, *b* will be the cost of the plant itself. The plant will cost money even if it is never used, just by buying the machinery. For each unit produced, there will be a cost for the labor and the material. Let *m* represent this cost to produce one unit of the product.

For a more concrete example, suppose a computer factory costs $100,000. It requires $100 of parts and $50 of labor to make one computer. How much will it cost for a company to make 1000 computers? To figure this, let *y* be the amount of money the company spends, and let *x* be the number of computers. The cost of the factory is $100,000, so $b = 100{,}000$. On the other hand, the cost of producing a computer is the parts plus labor, or $150, so $m = 150$. Therefore, $y = 150x + 100{,}000$. Substitute 1000 for *x* and get $y = 150 \times 1000 + 100{,}000 = 150{,}000 + 1000 = 250{,}000$. It will cost the company $250,000 to make 1000 computers.

Probabilities

Given a set of possible outcomes *X*, a *probability distribution* on *X* is a function that assigns a probability to each possible outcome. If the outcomes are $(x_1, x_2, x_3, \dots x_n)$, and the probability distribution is *p*, then the following rules are applied.

- $0 \leq p(x_i) \leq 1$, for any i.
- $\sum_{i=1}^{n} p(x_i) = 1$.

In other words, the probability of a given outcome must be between zero and 1, while the total probability must be 1.

If $p(x_i)$ is constant, then this is called a *uniform probability distribution*, and $p(x_i) = \frac{1}{n}$. For example, on a six-sided die, the probability of each of the six outcomes will be $\frac{1}{6}$.

If seeking the probability of an outcome occurring in some specific range *A* of possible outcomes, written $P(A)$, add up the probabilities for each outcome in that range. For example, consider a six-sided die, and figure the probability of getting a 3 or lower when it is rolled. The possible rolls are 1, 2, 3, 4, 5, and 6. So, to get a 3 or lower, a roll of 1, 2, or 3 must be completed. The probabilities of each of these is $\frac{1}{6}$, so add these to get $p(1) + p(2) + p(3) = \frac{1}{6} + \frac{1}{6} + \frac{1}{6} = \frac{1}{2}$.

Conditional Probabilities

An outcome occasionally lies within some range of possibilities *B*, and the probability that the outcomes also lie within some set of possibilities *A* needs to be figured. This is called a *conditional probability*. It is written as $P(A|B)$, which is read "the probability of *A* given *B*." The general formula for computing conditional probabilities is $P(A|B) = \frac{P(A \cap B)}{P(B)}$.

However, when dealing with uniform probability distributions, simplify this a bit. Write $|A|$ to indicate the number of outcomes in *A*. Then, for uniform probability distributions, write $P(A|B) = \frac{|A \cap B|}{|B|}$ (recall that $A \cap B$ means "*A* intersect *B*," and consists of all of the outcomes that lie in both *A* and *B*). This means that all possible outcomes do not need to be known. To see why this formula works, suppose that the set of outcomes *X* is $(x_1, x_2, x_3, \dots x_n)$, so that $|X| = n$. Then, for a uniform probability distribution, $P(A) = \frac{|A|}{n}$. However, this means:

$$(A|B) = \frac{P(A \cap B)}{P(B)} = \frac{\frac{|A \cap B|}{n}}{\frac{|B|}{n}} = \frac{|A \cap B|}{|B|}$$

since the *n*'s cancel out.

For example, suppose a die is rolled, and it is known that it will land between 1 and 4. However, how many sides the die has is unknown. Figure the probability that the die is rolled higher than 2. To figure this, $P(3)$ or $P(4)$ does not need to be determined, or any of the other probabilities, since it is known that a fair die has a uniform probability distribution. Therefore, apply the formula $\frac{|A \cap B|}{|B|}$. So, in this case *B* is (1, 2, 3, 4) and $A \cap B$ is (3, 4). Therefore $\frac{|A \cap B|}{|B|} = \frac{2}{4} = \frac{1}{2}$.

Conditional probability is an important concept because, in many situations, the likelihood of one outcome can differ radically depending on how something else comes out. The probability of passing a test given that one has studied all of the material is generally much higher than the probability of passing a test given that one has not studied at all. The probability of a person having heart trouble is much lower if that person exercises regularly. The probability that a college student will graduate is higher when his or her SAT scores are higher, and so on. For this reason, there are many people who are interested in conditional probabilities.

Note that in some practical situations, changing the order of the conditional probabilities can make the outcome very different. For example, the probability that a person with heart trouble has exercised regularly is quite different than the probability that a person who exercises regularly will have heart trouble. The probability of a person receiving a military-only award, given that he or she is or was a soldier, is generally not very high, but the probability that a person being or having been a soldier, given that he or she received a military-only award, is 1.

However, in some cases, the outcomes do not influence one another this way. If the probability of *A* is the same regardless of whether *B* is given; that is, if $P(A|B) = P(A)$, then *A* and *B* are considered *independent*. In this case, $P(A|B) = \frac{P(A \cap B)}{P(B)} = P(A)$, so $P(A \cap B) = P(A)P(B)$. In fact, if $P(A \cap B) = P(A)P(B)$, it can be determined that $P(A|B) = P(A)$ and $P(A|B) = P(B)$ by working backward. Therefore, *B* is also independent of *A*.

An example of something being independent can be seen in rolling dice. In this case, consider a red die and a green die. It is expected that when the dice are rolled, the outcome of the green die should not depend in any way on the outcome of the red die. Or, to take another example, if the same die is rolled repeatedly, then the next number rolled should not depend on which numbers have been rolled previously. Similarly, if a coin is flipped, then the next flip's outcome does not depend on the outcomes of previous flips.

This can sometimes be counter-intuitive, since when rolling a die or flipping a coin, there can be a streak of surprising results. If, however, it is known that the die or coin is fair, then these results are just the result of the fact that over long periods of time, it is very likely that some unlikely streaks of outcomes will occur. Therefore, avoid making the mistake of thinking that when considering a series of independent outcomes, a particular outcome is "due to happen" simply because a surprising series of outcomes has already been seen.

There is a second type of common mistake that people tend to make when reasoning about statistical outcomes: the idea that when something of low probability happens, this is surprising. It would be surprising that something with low probability happened after just one attempt. However, with so much happening all at once, it is easy to see at least something happen in a way that seems to have a very low probability. In fact, a lottery is a good example. The odds of winning a lottery are very small, but the odds that somebody wins the lottery each week are actually fairly high. Therefore, no one should be surprised when some low probability things happen.

Addition Rule

The *addition rule* for probabilities states that the probability of *A* or *B* happening is $P(A \cup B) = P(A) + P(B) - P(A \cap B)$. Note that the subtraction of $P(A \cap B)$ must be performed, or else it would result in double counting any outcomes that lie in both *A* and in *B*. For example, suppose that a 20-sided die is being rolled. Fred bets that the outcome will be greater than 10, while Helen bets that it will be greater than 4 but less than 15. What is the probability that at least one of them is correct?

We apply the rule $P(A \cup B) = P(A) + P(B) - P(A \cap B)$, where *A* is that outcome *x* is in the range $x > 10$, and *B* is that outcome *x* is in the range $4 < x < 15$.

$$P(A) = 10 \times \frac{1}{20} = \frac{1}{2}$$

$$P(B) = 10 \times \frac{1}{20} = \frac{1}{2}$$

$P(A \cap B)$ can be computed by noting that $A \cap B$ means the outcome x is in the range $10 < x < 15$, so $P(A \cap B) = 4 \times \frac{1}{20} = \frac{1}{5}$. Therefore:

$$P(A \cup B) = P(A) + P(B) - P(A \cap B)$$

$$\frac{1}{2} + \frac{1}{2} - \frac{1}{5} = \frac{4}{5}$$

Note that in this particular example, we could also have directly reasoned about the set of possible outcomes $A \cup B$, by noting that this would mean that x must be in the range $5 \leq x$. However, this is not always the case, depending on the given information.

Multiplication Rule

The *multiplication rule* for probabilities states the probability of A and B both happening is $P(A \cap B) = P(A)P(B|A)$. As an example, suppose that when Jamie wears black pants, there is a ½ probability that she wears a black shirt as well, and that she wears black pants ¾ of the time. What is the probability that she is wearing both a black shirt and black pants?

To figure this, use the above formula, where A will be "Jamie is wearing black pants," while B will be "Jamie is wearing a black shirt." It is known that $P(A)$ is ¾. It is also known that $P(B|A) = \frac{1}{2}$. Multiplying the two, the probability that she is wearing both black pants and a black shirt is:

$$P(A)P(B|A) = \frac{3}{4} \times \frac{1}{2} = \frac{3}{8}$$

Complex Numbers

Some types of equations can be solved to find real answers, but this is not the case for all equations. For example, $x^2 = k$ can be solved when k is non-negative, but it has no real solutions when k is negative. Equations do have solutions if complex numbers are allowed.

Complex numbers are defined in the following manner: every complex number can be written as $a + bi$, where $i^2 = -1$. Thus, the solutions to the equation $x^2 = -1$ are $\pm i$.

In order to find roots of negative numbers more generally, the properties of roots (or of exponents) are used. For example, $\sqrt{-4} = \sqrt{-1}\sqrt{4} = \pm 2i$. All arithmetic operations can be performed with complex numbers, where i is like any other constant. The value of i^2 can replace -1.

Series and Sequences

A *sequence* is an infinite, ordered list of numbers and a function whose domain is an infinite subset of non-negative integers. Usually, the domain will be all non-negative integers, or else all positive integers, but sometimes, it is convenient to use a smaller subset for some sequences. Although it would be possible to use function notation, it is customary to write sequences a little bit differently.

Given a sequence defined by some function, the value of this function is written on n as a_n and denoted in the entire sequence as $\{a_n\}$. The sequence can be written $a_1, a_2, a_3, \ldots a_n, \ldots$, keeping in mind this list continues infinitely. The a_n values are called the *terms* of the sequence.

In some cases, a formula for a_n is an expression that only involves *n* and constants. In some other cases, an expression for a_n involves only *n*, constants, and a_{n-1}. This latter case is called a *recursive* definition for the sequences.

An *infinite sum* or simply *sum* of a sequence $\{a_n\}$ is a sequence $\{s_n\}$, where $s_n = a_1 + a_2 + \cdots + a_n$. Some sequences have the property of getting closer to one particular value. The value is the *limit* of the sequence.

If the limit of a sequence $\{a_n\}$ is *L*, this means that for any positive real number δ, there is a value of *M*, as long as $n > M$, $|a_n - L| < \delta$. This is just a very formal way of saying that for any real number positive (real number δ), there is a point where all the remaining values are within δ of *L* in the sequence. This just means, on the whole, getting closer to *L* as the sequence ends. This is a *limit* of a sequence $\lim_{n\to\infty} a_n$.

If a sequence has a limit, the sequence *converges*. On the other hand, if the absolute value $|a_n|$ keeps on getting bigger, the sequence *diverges.*

Some sequences do not converge or diverge. For example, the sequence $a_n = \begin{cases} 1, n \text{ odd} \\ -1, n \text{ even} \end{cases}$, flips back and forth between 1 and -1. This sequence never converges, since it keeps bouncing back and forth. However, it does not diverge, either, since the absolute value is never bigger than 1.

In order to find the limit of a sequence, we can use the following rules:

- $\lim_{n\to\infty} k = k$ for all real numbers *k*.
- $\lim_{n\to\infty} \frac{1}{n} = 0.$
- $\lim_{n\to\infty} n = \infty.$
- $\lim_{n\to\infty} \frac{k}{n^p} = 0$ when *k* is real and *p* is a positive rational number.
- $\lim_{n\to\infty} (a_n + b_n) = \lim_{n\to\infty} a_n + \lim_{n\to\infty} b_n$
- $\lim_{n\to\infty} (a_n - b_n) = \lim_{n\to\infty} a_n - \lim_{n\to\infty} b_n$
- $\lim_{n\to\infty} (a_n \times b_n) = \lim_{n\to\infty} a_n \times \lim_{n\to\infty} b_n$. As a special case, $\lim_{n\to\infty} ka_n = k \lim_{n\to\infty} a_n$
- $\lim_{n\to\infty} \left(\frac{a_n}{b_n}\right) = \frac{\lim_{n\to\infty} a_n}{\lim_{n\to\infty} b_n}$ when $\lim_{n\to\infty} b_n$ is not 0.

A sequence is called *monotonic* if the terms of the sequence never decrease or if they never increase. In other words, $\{a_n\}$ is monotonic if one of two things happen: either $a_n \geq a_m$ whenever $n > m$ (in this case, the sequences is also called *non-decreasing*), or else $a_n \leq a_m$ whenever $n > m$ (in this case, the sequence is also called *non-increasing*).

A sequence is said to be *bounded above* by *k* if, for any value of *n*, every $a_n \leq k$, and *bounded below* by *k* if $a_n \geq k$. Every non-decreasing sequence that is bounded above converges to some real number. Every non-increasing sequence that is bounded below converges to some real number.

An *arithmetic sequence* is a sequence where the next term is obtained from the previous term by adding a specific quantity, *k*. In other words, $a_{n+1} = a_n + k$. Another way of writing this out is the sequence $a_1, a_1 + k, a_1 + 2k, \dots, a_1 + (n-1)k$ .... That is, $a_n = a_1 + (n-1)k$. The *sum* of the first *n* terms of an arithmetic sequence is $s_n = \frac{n}{2}(a_1 + a_n)$.

A *geometric sequence* (or a *geometric progression*) is a sequence in which for some specific quantity *r*, $a_{n+1} = ra_n$. Another way of writing this is that the sequence is $a_1, a_1r, a_1r^2, \dots, a_1r^{n-1}$ .... The general formula is $a_n = a_1r^{n-1}$. The sum of the first *n* terms of a geometric sequence is:

$$s_n = \frac{a_1(1-r^n)}{1-r}$$

A *series* or *infinite series* is a sequence $\{s_n\}$, whose *n*-th term is the sum of the first *n* terms of some sequence $\{a_n\}$. Thus, $s_n = a_1 + a_2 + \cdots + a_n$ can also be written as $\sum_{m=1}^{n} a_m$.

Each s_n is the sum of the first n terms and is called the *n-th partial sum.* The *infinite sum* (or simply, *sum*) of a sequence $\{a_n\}$ is the limit of the sequence $\{s_n\}$, also written as $\sum_{n=1}^{\infty} a_n$.

A series can converge, diverge, or neither, just like a sequence. The rules for finding whether or not a series converges or diverges are identical to the rules for sequences. When a series is being added or subtracted, or multiplied by constants, it obeys the following rules:

$$\sum_{n=1}^{\infty}(a_n + b_n) = \sum_{n=1}^{\infty} a_n + \sum_{n=1}^{\infty} b_n$$

$$\sum_{n=1}^{\infty}(a_n - b_n) = \sum_{n=1}^{\infty} a_n - \sum_{n=1}^{\infty} b_n$$

$$\sum_{n=1}^{\infty} ka_n = k\sum_{n=1}^{\infty} a_n$$

A *geometric series* is a sum of a geometric sequence:

$$\sum_{n=1}^{\infty} ar^{n-1} = a_1 + a_2r + \cdots + a_nr^{n-1} + \cdots$$

In such a series, when $|r| \geq 1$, the series diverges. However, when $|r| < 1$, then:

$$\sum_{n=1}^{\infty} ar^{n-1} = \frac{a}{1-r}$$

It's important to note that whenever a sum $\sum_{n=1}^{\infty} a_n$ converges, the sequence $\{a_n\}$ has a limit of 0: $\lim_{n\to\infty} a_n = 0$. This is one possible test to see whether or not a series converges. However, just because this limit is zero does not mean that the sum diverges, so the test only works in one direction.

Determinants

A *matrix* is a rectangular arrangement of numbers in rows and columns. The *determinant* of a matrix is a special value that can be calculated for any square matrix.

Using the *square 2 x 2 matrix* $\begin{bmatrix} a & b \\ c & d \end{bmatrix}$, the determinant is $ad - bc$.

For example, the determinant of the matrix $\begin{bmatrix} -5 & 1 \\ 3 & 4 \end{bmatrix}$ is *-5(4) – 1(3) = -20 – 3 = -23.*

Using a *3 x 3 matrix* $\begin{bmatrix} a & b & c \\ d & e & f \\ g & h & i \end{bmatrix}$, the determinant is $a(ei - fh) - b(di - fg) + c(dh - eg)$.

For example, the determinant of the matrix $\begin{bmatrix} 2 & 0 & 1 \\ -1 & 3 & 2 \\ 2 & -2 & -1 \end{bmatrix}$ is

$$2\big(3(-1) - 2(-2)\big) - 0\big(-1(-1) - 2(2)\big) + 1\big(-1(-2) - 3(2)\big)$$

$$= 2(-3 + 4) - 0(1 - 4) + 1(2 - 6)$$

$$= 2(1) - 0(-3) + 1(-4)$$

$$= 2 - 0 - 4 = -2$$

The pattern continues for larger square matrices.

Permutations and Combinations

The *factorial* is defined for non-negative integers. The factorial of *n* is written as $n!$ For positive integers, the factorial is defined as the product of all positive integers up to *n*. So, $n! = 1 \times 2 \times \ldots \times n$.

For zero, $0! = 1$. The reason for zero being a special case is two-fold. First, the relation is always $n! = n \times (n-1)!$ when the right hand side is defined. Second, it makes the choice functions below work out correctly.

The *combinatorial choice function* indicates how many distinct ways one can choose to pick out *k* objects from a set of *n* objects. The choice function is written as $\binom{n}{k}$—read as "*n* choose *k*"—and is given by $\binom{n}{k} = \frac{n!}{k! \times (n-k)!}$.

As an example, suppose a person wanted to choose three shirts out of five shirts to take on a trip. How many ways can this be done?

The answer is given by computing 5 choose 3.

$$\binom{5}{3} = \frac{5!}{3! \times (5-3)!} = \frac{5 \times 4 \times 3 \times 2 \times 1}{3 \times 2 \times 1 \times 2 \times 1}$$

At this stage, all the indicated multiplications can be calculated, but a number of common factors cancel first. (The appearance of many common factors is normal when computing choice functions.)

The expression simplifies to:

$$\frac{5 \times 4 \times 3 \times 2 \times 1}{3 \times 2 \times 1 \times 2 \times 1} = \frac{5 \times 4}{2 \times 1} = \frac{20}{2} = 10$$

There are ten different combinations of three shirts.

Using the choice function, it can be calculated how many ways to order a set of objects. Ordering a set of n objects requires choosing a first object out of n objects, then picking a second object out of the remaining $n - 1$ objects. Then a third object out of the $n - 2$ objects can be picked, and so on. Therefore, the total number of ways to order n objects will be $\binom{n}{1}\binom{n-1}{1} \dots \binom{1}{1}$

Practice Questions

1. 3.4 + 2.35 + 4 =
   a. 5.35
   b. 9.2
   c. 9.75
   d. 10.25

2. $5.88 \times 3.2 =$
   a. 18.816
   b. 16.44
   c. 20.352
   d. 17

3. $\frac{3}{25} =$
   a. 0.15
   b. 0.1
   c. 0.9
   d. 0.12

4. Which of the following is largest?
   a. 0.45
   b. 0.096
   c. 0.3
   d. 0.313

5. Which of the following is NOT a way to write 40 percent of *N*?
   a. $(0.4)N$
   b. $\frac{2}{5}N$
   c. $40N$
   d. $\frac{4N}{10}$

6. Which is closest to 17.8×9.9?
   a. 140
   b. 180
   c. 200
   d. 350

7. A student gets an 85% on a test with 20 questions. How many answers did the student solve correctly?
   a. 15
   b. 16
   c. 17
   d. 18

8. Four people split a bill. The first person pays for $\frac{1}{5}$, the second person pays for $\frac{1}{4}$, and the third person pays for $\frac{1}{3}$. What fraction of the bill does the fourth person pay?

a. $\frac{13}{60}$
b. $\frac{47}{60}$
c. $\frac{1}{4}$
d. $\frac{4}{15}$

9. 6 is 30% of what number?

a. 18
b. 20
c. 24
d. 26

10. $3\frac{2}{3} - 1\frac{4}{5} =$

a. $1\frac{13}{15}$
b. $\frac{14}{15}$
c. $2\frac{2}{3}$
d. $\frac{4}{5}$

11. What is $\frac{420}{98}$ rounded to the nearest integer?

a. 4
b. 3
c. 5
d. 6

12. $4\frac{1}{3} + 3\frac{3}{4} =$

a. $6\frac{5}{12}$
b. $8\frac{1}{12}$
c. $8\frac{2}{3}$
d. $7\frac{7}{12}$

13. Five of six numbers have a sum of 25. The average of all six numbers is 6. What is the sixth number?

a. 8
b. 10
c. 11
d. 12

14. $52.3 \times 10^{-3} =$
   a. 0.00523
   b. 0.0523
   c. 0.523
   d. 523

15. If $\frac{5}{2} \div \frac{1}{3} = n$, then n is between:
   a. 5 and 7
   b. 7 and 9
   c. 9 and 11
   d. 3 and 5

16. A closet is filled with red, blue, and green shirts. If $\frac{1}{3}$ of the shirts are green and $\frac{2}{5}$ are red, what fraction of the shirts are blue?
   a. $\frac{4}{15}$
   b. $\frac{1}{5}$
   c. $\frac{7}{15}$
   d. $\frac{1}{2}$

17. Shawna buys $2\frac{1}{2}$ gallons of paint. If she uses $\frac{1}{3}$ of it on the first day, how much does she have left?
   a. $1\frac{5}{6}$ gallons
   b. $1\frac{1}{2}$ gallons
   c. $1\frac{2}{3}$gallons
   d. 2 gallons

18. What is the value of $x^2 - 2xy + 2y^2$ when $x = 2, y = 3$?
   a. 8
   b. 10
   c. 12
   d. 14

19. $(2x - 4y)^2 =$
   a. $4x^2 - 16xy + 16y^2$
   b. $4x^2 - 8xy + 16y^2$
   c. $4x^2 - 16xy - 16y^2$
   d. $2x^2 - 8xy + 8y^2$

20. If $x > 3$, then $\frac{x^2-6x+9}{x^2-x-6} =$

a. $\frac{x+2}{x-3}$
b. $\frac{x-2}{x-3}$
c. $\frac{x-3}{x+3}$
d. $\frac{x-3}{x+2}$

21. The square and circle have the same center. The circle has a radius of *r*. What is the area of the shaded region?

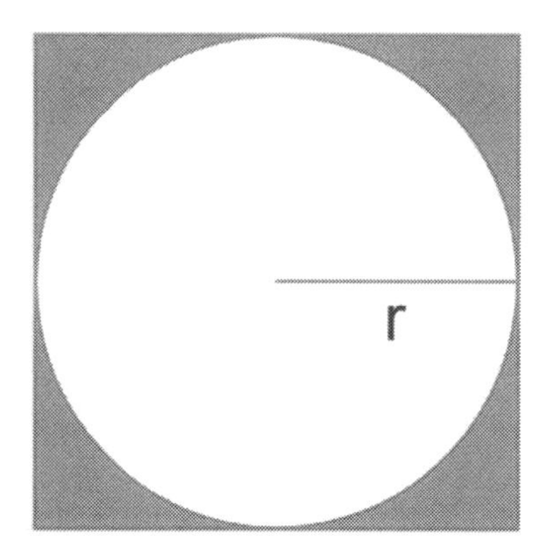

a. $r^2 - \pi r^2$
b. $4r^2 - 2\pi r$
c. $(4 - \pi)r^2$
d. $(\pi - 1)r^2$

22. If $4x - 3 = 5$, then $x =$

a. 1
b. 2
c. 3
d. 4

23. Solve for *x*, if $x^2 - 2x - 8 = 0$.

a. $2 \pm \frac{\sqrt{30}}{2}$
b. $2 \pm 4\sqrt{2}$
c. 1 ± 3
d. $4 \pm \sqrt{2}$

24. Which of the following is a factor of both $x^2 + 4x + 4$ and $x^2 - x - 6$?

a. $x - 3$
b. $x + 2$
c. $x - 2$
d. $x + 3$

25. Write the expression for three times the sum of twice a number and one minus 6.

a. $2x + 1 - 6$
b. $3x + 1 - 6$
c. $3(x + 1) - 6$
d. $3(2x + 1) - 6$

26. On Monday, Robert mopped the floor in 4 hours. On Tuesday, he did it in 3 hours. If on Monday, his average rate of mopping was *p* sq. ft. per hour, what was his average rate on Tuesday?

a. $\frac{4}{3}p$ sq. ft. per hour
b. $\frac{3}{4}p$ sq. ft. per hour
c. $\frac{5}{4}p$ sq. ft. per hour
d. $p + 1$ sq. ft. per hour

27. Which of the following inequalities is equivalent to $3 - \frac{1}{2}x \geq 2$?

a. $x \geq 2$
b. $x \leq 2$
c. $x \geq 1$
d. $x \leq 1$

28. For which of the following are $x = 4$ and $x = -4$ solutions?

a. $x^2 + 16 = 0$
b. $x^2 + 4x - 4 = 0$
c. $x^2 - 2x - 2 = 0$
d. $x^2 - 16 = 0$

29. If *x* is not zero, then $\frac{3}{x} + \frac{5u}{2x} - \frac{u}{4} =$

a. $\frac{12+10u-ux}{4x}$
b. $\frac{3+5u-ux}{x}$
c. $\frac{12x+10u+ux}{4x}$
d. $\frac{12+10u-u}{4x}$

30. If $6t + 4 = 16$, what is *t*?

a. 1
b. 2
c. 3
d. 4

31. The variable *y* is directly proportional to *x*. If $y = 3$ when $x = 5$, then what is *y* when $x = 20$?

a. 10
b. 12
c. 14
d. 16

32. A line passes through the point (1, 2) and crosses the *y*-axis at *y* = 1. Which of the following is an equation for this line?

a. $y = 2x$
b. $y = x + 1$
c. $x + y = 1$
d. $y = \frac{x}{2} - 2$

33. There are 4*x* + 1 treats in each party favor bag. If a total of 60*x* + 15 treats are distributed, how many bags are given out?

a. 15
b. 16
c. 20
d. 22

34. Apples cost $2 each, while bananas cost $3 each. Maria purchased 10 fruits in total and spent $22. How many apples did she buy?

a. 5
b. 6
c. 7
d. 8

35. What are the polynomial roots of $x^2 + x - 2$?

a. 1 and -2
b. -1 and 2
c. 2 and -2
d. 9 and 13

36. What is the *y*-intercept of $y = x^{5/3} + (x - 3)(x + 1)$?

a. 3.5
b. 7.6
c. -3
d. -15.1

37. $x^4 - 16$ can be simplified to which of the following?

a. $(x^2 - 4)(x^2 + 4)$
b. $(x^2 + 4)(x^2 + 4)$
c. $(x^2 - 4)(x^2 - 4)$
d. $(x^2 - 2)(x^2 + 4)$

38. $(4x^2y^4)^{\frac{3}{2}}$ can be simplified to which of the following?

a. $8x^3y^6$
b. $4x^{\frac{5}{2}}y$
c. $4xy$
d. $32x^{\frac{7}{2}}y^{\frac{11}{2}}$

39. If $\sqrt{1+x} = 4$, what is *x*?
   a. 10
   b. 15
   c. 20
   d. 25

40. Suppose $\frac{x+2}{x} = 2$. What is *x*?
   a. -1
   b. 0
   c. 2
   d. 4

41. A ball is thrown from the top of a high hill, so that the height of the ball as a function of time is $h(t) = -16t^2 + 4t + 6$, in feet. What is the maximum height of the ball in feet?
   a. 6
   b. 6.25
   c. 6.5
   d. 6.75

42. A rectangle has a length that is 5 feet longer than three times its width. If the perimeter is 90 feet, what is the length in feet?
   a. 10
   b. 20
   c. 25
   d. 35

43. Five students take a test. The scores of the first four students are 80, 85, 75, and 60. If the median score is 80, which of the following could NOT be the score of the fifth student?
   a. 60
   b. 80
   c. 85
   d. 100

44. In an office, there are 50 workers. A total of 60% of the workers are women, and the chances of a woman wearing a skirt is 50%. If no men wear skirts, how many workers are wearing skirts?
   a. 12
   b. 15
   c. 16
   d. 20

45. Ten students take a test. Five students get a 50. Four students get a 70. If the average score is 55, what was the last student's score?
   a. 20
   b. 40
   c. 50
   d. 60

46. A company invests $50,000 in a building where they can produce saws. If the cost of producing one saw is $40, then which function expresses the amount of money the company pays? The variable *y* is the money paid and *x* is the number of saws produced.

a. $y = 50{,}000x + 40$
b. $y + 40 = x - 50{,}000$
c. $y = 40x - 50{,}000$
d. $y = 40x + 50{,}000$

47. A six-sided die is rolled. What is the probability that the roll is 1 or 2?

a. $\frac{1}{6}$
b. $\frac{1}{4}$
c. $\frac{1}{3}$
d. $\frac{1}{2}$

48. A line passes through the origin and through the point (-3, 4). What is the slope of the line?

a. $-\frac{4}{3}$
b. $-\frac{3}{4}$
c. $\frac{4}{3}$
d. $\frac{3}{4}$

49. An equilateral triangle has a perimeter of 18 feet. If a square whose sides have the same length as one side of the triangle is built, what will be the area of the square?

a. 6 square feet
b. 36 square feet
c. 256 square feet
d. 1000 square feet

50. Find the determinant of the matrix $\begin{bmatrix} -4 & 2 \\ 3 & -1 \end{bmatrix}$.

a. -10
b. -2
c. 0
d. 2

51. If $a \neq b$, solve for x if $\frac{1}{x} + \frac{2}{a} = \frac{2}{b}$

a. $\frac{a-b}{ab}$
b. $\frac{ab}{2(a-b)}$
c. $\frac{2(a-b)}{ab}$
d. $\frac{a-b}{2ab}$

52. If $x^2 + x - 3 = 0$, then $\left(x + \frac{1}{2}\right)^2 =$

a. $\frac{11}{2}$

b. $\frac{13}{4}$

c. 11

d. $\frac{121}{4}$

53. Which graph will be a line parallel to the graph of $y = 3x - 2$?

a. $2y - 6x = 2$
b. $y - 4x = 4$
c. $3y = x - 2$
d. $2x - 2y = 2$

54. An equation for the line passing through the origin and the point $(2, 1)$ is

a. $y = 2x$
b. $y = \frac{1}{2}x$
c. $y = x - 2$
d. $2y = x + 1$

55. Jessica buys 10 cans of paint. Red paint costs $1 per can and blue paint costs $2 per can. In total, she spends $16. How many red cans did she buy?

a. 2
b. 3
c. 4
d. 5

56. A farmer owns two (non-adjacent) square plots of land, which he wishes to fence. The area of one is 1000 square feet, while the area of the other is 10 square feet. How much fencing does he need, in feet?

a. 44
b. $40\sqrt{10}$
c. $440\sqrt{10}$
d. $44\sqrt{10}$

57. If $\log_{10} x = 2$, then *x* is

a. 4
b. 20
c. 100
d. 1000

58. Let $f(x) = 2x + 1, g(x) = \frac{x-1}{4}$. Find $g(f(x))$.

a. $\frac{x+1}{2}$

b. $\frac{x}{2}$

c. $\frac{2x^2-x-1}{4}$

d. $3x$

59. Suppose θ is an acute angle, and $\sin\theta = \frac{\sqrt{3}}{2}$. What is $\cos\theta$?

a. $\frac{1}{2}$
b. $\frac{\sqrt{3}}{2}$
c. $\frac{\sqrt{2}}{2}$
d. $\frac{1}{4}$

60. $2x(3x+1) - 5(3x+1) =$

a. $10x(3x+1)$
b. $10x^2(3x+1)$
c. $(2x-5)(3x+1)$
d. $(2x+1)(3x-5)$

61. For which real numbers *x* is $-3x^2 + x - 8 > 0$?

a. All real numbers *x*
b. $-2\sqrt{\frac{2}{3}} < x < 2\sqrt{\frac{2}{3}}$
c. $1 - 2\sqrt{\frac{2}{3}} < x < 1 + 2\sqrt{\frac{2}{3}}$
d. For no real numbers *x*

62. A root of $x^2 - 2x - 2$ is

a. $1 + \sqrt{3}$
b. $1 + 2\sqrt{2}$
c. $2 + 2\sqrt{3}$
d. $2 - 2\sqrt{3}$

63. In the *xy*-plane, the graph of $y = x^2 + 2$ and the circle with center $(0,1)$ and radius 1 have how many points of intersection?

a. 0
b. 1
c. 2
d. 3

64. A line goes through the point (-4, 0) and the point (0,2). What is the slope of the line?

a. 2
b. 4
c. $\frac{3}{2}$
d. $\frac{1}{2}$

65. How many different ways can we order the letters *a*, *b*, *c*?

a. 3

b. 6

c. 9

d. 12

66. If $f(x) = 4x + 2$, and $f^{-1}(x)$ is the inverse function for *f*, then what is $f^{-1}(6)$?

a. 0

b. $\frac{1}{2}$

c. 1

d. $\frac{3}{2}$

67. The sequence $\{a_n\}$ is defined by the relation $a_{n+1} = 3a_n - 1, a_1 = 1$. Find a_3.

a. 2

b. 4

c. 5

d. 15

68. Six people apply to work for Janice's company, but she only needs four workers. How many different groups of four employees can Janice choose?

a. 6

b. 10

c. 15

d. 36

69. If $f(x) = (\frac{1}{2})^x$ and $a < b$, then which of the following must be true?

a. $f(a) < f(b)$

b. $f(a) > f(b)$

c. $f(a) + f(b) = 0$

d $3f(a) = f(b)$

Answer Explanations

1. C: The decimal points are lined up, with zeroes put in as needed. Then, the numbers are added just like integers:

$$\begin{array}{r} 3.40 \\ 2.35 \\ +\underline{4.00} \\ 9.75 \end{array}$$

2. A: This problem can be multiplied as 588×32, except at the end, the decimal point needs to be moved three places to the left. Performing the multiplication will give 18,816, and moving the decimal place over three places results in 18.816.

3. D: The fraction is converted so that the denominator is 100 by multiplying the numerator and denominator by 4, to get $\frac{3}{25} = \frac{12}{100}$. Dividing a number by 100 just moves the decimal point two places to the left, with a result of 0.12.

4. A: Figure out which is largest by looking at the first non-zero digits. Choice *B*'s first non-zero digit is in the hundredths place. The other three all have non-zero digits in the tenths place, so it must be *A*, *C*, or *D*. Of these, *A* has the largest first non-zero digit.

5. C: 40*N* would be 4000% of *N*. It's possible to check that each of the others is actually 40% of *N*.

6. B: Instead of multiplying these out, the product can be estimated by using $18 \times 10 = 180$. The error here should be lower than 15, since it is rounded to the nearest integer, and the numbers add to something less than 30.

7. C: 85% of a number means multiplying that number by 0.85. So, $0.85 \times 20 = \frac{85}{100} \times \frac{20}{1}$, which can be simplified to $\frac{17}{20} \times \frac{20}{1} = 17$.

8. A: To find the fraction of the bill that the first three people pay, the fractions need to be added, which means finding common denominator. The common denominator will be 60. $\frac{1}{5} + \frac{1}{4} + \frac{1}{3} = \frac{12}{60} + \frac{15}{60} + \frac{20}{60} = \frac{47}{60}$. The remainder of the bill is $1 - \frac{47}{60} = \frac{60}{60} - \frac{47}{60} = \frac{13}{60}$.

9. B: 30% is 3/10. The number itself must be 10/3 of 6, or $\frac{10}{3} \times 6 = 10 \times 2 = 20$.

10. A: First, convert the mixed numbers to improper fractions: $\frac{11}{3} - \frac{9}{5}$. Then, use 15 as a common denominator: $\frac{11}{3} - \frac{9}{5} =: \frac{55}{15} - \frac{27}{15} = \frac{28}{15} = 1\frac{13}{15}$ (when rewritten to get rid of the improper fraction).

11. A: Dividing by 98 can be approximated by dividing by 100, which would mean shifting the decimal point of the numerator to the left by 2. The result is 4.2 and rounds to 4.

12. B: $4\frac{1}{3} + 3\frac{3}{4} = 4 + 3 + \frac{1}{3} + \frac{3}{4} = 7 + \frac{1}{3} + \frac{3}{4}$. Adding the fractions gives $\frac{1}{3} + \frac{3}{4} = \frac{4}{12} + \frac{9}{12} = \frac{13}{12} = 1 + \frac{1}{12}$. Thus, $7 + \frac{1}{3} + \frac{3}{4} = 7 + 1 + \frac{1}{12} = 8\frac{1}{12}$.

13. C: The average is calculated by adding all six numbers, then dividing by 6. The first five numbers have a sum of 25. If the total divided by 6 is equal to 6, then the total itself must be 36. The sixth number must be 36 – 25 = 11.

14. B: Multiplying by 10^{-3} means moving the decimal point three places to the left, putting in zeroes as necessary.

15. B: $\frac{5}{2} \div \frac{1}{3} = \frac{5}{2} \times \frac{3}{1} = \frac{15}{2} = 7.5$.

16. A: The total fraction taken up by green and red shirts will be $\frac{1}{3} + \frac{2}{5} = \frac{5}{15} + \frac{6}{15} = \frac{11}{15}$. The remaining fraction is $1 - \frac{11}{15} = \frac{15}{15} - \frac{11}{15} = \frac{4}{15}$.

17. C: If she has used 1/3 of the paint, she has 2/3 remaining. $2\frac{1}{2}$ gallons are the same as $\frac{5}{2}$ gallons. The calculation is $\frac{2}{3} \times \frac{5}{2} = \frac{5}{3} = 1\frac{2}{3}$ gallons.

18. B: Start with the original equation: x- 2xy + 2y, then replace each instance of x with a 2, and each instance of y with a 3 to get:

$$2^2 - 2 \times 2 \times 3 + 2 \times 3^2$$
$$4 - 12 + 18 = 10$$

19. A: To expand a squared binomial, it's necessary to use the *First, Inner, Outer, Last Method.*

$$(2x - 4y)^2$$

$$(2x)(2x) + (2x)(-4y) + (-4y)(2x) + (-4y)(-4y)$$

$$4x^2 - 8xy - 8xy + 16y^2$$

$$4x^2 - 16xy + 16y^2$$

20. D: Factor the numerator into $x^2 - 6x + 9 = (x - 3)^2$, since $-3 - 3 = -6, (-3)(-3) = 9$. Factor the denominator into $x^2 - x - 6 = (x - 3)(x + 2)$, since $-3 + 2 = -1, (-3)(2) = -6$. This means the rational function can be rewritten as:

$$\frac{x^2 - 6x + 9}{x^2 - x - 6} = \frac{(x - 3)^2}{(x - 3)(x + 2)}$$

Using the restriction of x > 3, do not worry about any of these terms being 0, and cancel an $x - 3$ from the numerator and the denominator, leaving $\frac{x-3}{x+2}$.

21. C: The area of the shaded region is the area of the square, minus the area of the circle. The area of the circle will be πr^2. The side of the square will be $2r$, so the area of the square will be $4r^2$. Therefore, the difference is $4r^2 - \pi r^2 = (4 - \pi)r^2$.

22. B: Add 3 to both sides to get $4x = 8$. Then divide both sides by 4 to get $x = 2$.

23. C: The numbers needed are those that add to -2 and multiply to -8. The difference between 2 and 4 is 2. Their product is 8, and -4 and 2 will work. Therefore, $x^2 - 2x - 8 = (x - 4)(x + 2)$. The latter has roots 4 and -2 or 1 ± 3.

24. B: To factor $x^2 + 4x + 4$, the numbers needed are those that add to 4 and multiply to 4. Therefore, both numbers must be 2, and the expression factors to $x^2 + 4x + 4 = (x + 2)^2$. Similarly, the second expression factors to $x^2 - x - 6 = (x - 3)(x + 2)$, so that they have $x + 2$ in common.

25. D: The expression is three times the sum of twice a number and 1, which is $3(2x + 1)$. Then, 6 is subtracted from this expression.

26. A: Robert accomplished his task on Tuesday in ¾ the time compared to Monday. He must have worked 4/3 as fast.

27. B: To simplify this inequality, subtract 3 from both sides to get $-\frac{1}{2}x \geq -1$. Then, multiply both sides by -2 (remembering this flips the direction of the inequality) to get $x \leq 2$.

28. D: There are two ways to approach this problem. Each value can be substituted into each equation. A can be eliminated, since $4^2 + 16 = 32$. Choice *B* can be eliminated, since $4^2 + 4 \times 4 - 4 = 28$. *C* can be eliminated, since $4^2 - 2 \times 4 - 2 = 6$. But, plugging in either value into $x^2 - 16$, which gives $(\pm 4)^2 - 16 = 16 - 16 = 0$.

29. A: The common denominator here will be 4*x*. Rewrite these fractions as:

$$\frac{3}{x} + \frac{5u}{2x} - \frac{u}{4}$$

$$\frac{12}{4x} + \frac{10u}{4x} - \frac{ux}{4x}$$

$$\frac{12x + 10u - ux}{4x}$$

30. B: First, subtract 4 from each side. This yields $6t = 12$. Now, divide both sides by 6 to obtain $t = 2$.

31. B: To be directly proportional means that $y = mx$. If *x* is changed from 5 to 20, the value of *x* is multiplied by 4. Applying the same rule to the y-value, also multiply the value of *y* by 4. Therefore, *y* = 12.

32. B: From the slope-intercept form, $y = mx + b$, it is known that *b* is the *y*-intercept, which is 1. Compute the slope as $\frac{2-1}{1-0} = 1$, so the equation should be $y = x + 1$.

33. A: Each bag contributes 4*x* + 1 treats. The total treats will be in the form $4nx + n$ where *n* is the total number of bags. The total is in the form 60*x* + 15, from which it is known $n = 15$.

34. D: Let *a* be the number of apples and *b* the number of bananas. Then, the total cost is $2a + 3b = 22$, while it also known that $a + b = 10$. Using the knowledge of systems of equations, cancel the *b* variables by multiplying the second equation by -3. This makes the equation $-3a - 3b = -30$. Adding this to the first equation, the b values cancel to get $-a = -8$, which simplifies to *a* = 8.

35. A: Finding the roots means finding the values of *x* when *y* is zero. The quadratic formula could be used, but in this case it is possible to factor by hand, since the numbers -1 and 2 add to 1 and multiply to -2. So, factor $x^2 + x - 2 = (x - 1)(x + 2) = 0$, then set each factor equal to zero. Solving for each value gives the values *x* = 1 and *x* = -2.

36. C: To find the *y*-intercept, substitute zero for *x*, which gives us:

$$y = 0^{\frac{5}{3}} + (0 - 3)(0 + 1)$$

$$0 + (-3)(1) = -3$$

37. A: This has the form $t^2 - y^2$, with $t = x^2$ and $y = 4$. It's also known that $t^2 - y^2 = (t + y)(t - y)$, and substituting the values for *t* and *y* into the right-hand side gives $(x^2 - 4)(x^2 + 4)$.

38. A: Simplify this to:

$$(4x^2y^4)^{\frac{3}{2}} = 4^{\frac{3}{2}}(x^2)^{\frac{3}{2}}(y^4)^{\frac{3}{2}}$$

Now, $4^{\frac{3}{2}} = (\sqrt{4})^3 = 2^3 = 8$. For the other, recall that the exponents must be multiplied, so this yields:

$$8x^{2\times\frac{3}{2}}y^{4\times\frac{3}{2}} = 8x^3y^6$$

39. B: Start by squaring both sides to get $1 + x = 16$. Then subtract 1 from both sides to get $x = 15$.

40. C: Multiply both sides by *x* to get $x + 2 = 2x$, which simplifies to $-x = -2$, or *x* = 2.

41. B: The independent variable's coordinate at the vertex of a parabola (which is the highest point, when the coefficient of the squared independent variable is negative) is given by $x = -\frac{b}{2a}$. Substitute and solve for *x* to get $x = -\frac{4}{2(-16)} = \frac{1}{8}$. Using this value of *x*, the maximum height of the ball (*y*), can be calculated. Substituting *x* into the equation yields $h(t) = -16\frac{1}{8}^2 + 4\frac{1}{8} + 6 = 6.25$.

42. D: Denote the width as *w* and the length as *l*. Then, $l = 3w + 5$. The perimeter is $2w + 2l = 90$. Substituting the first expression for *l* into the second equation yields $2(3w + 5) + 2w = 90$, or $8w = 80$, so w $= 10$. Putting this into the first equation, it yields $l = 3(10) + 5 = 35$.

43. A: Lining up the given scores provides the following list: 60, 75, 80, 85, and one unknown. Because the median needs to be 80, it means 80 must be the middle data point out of these five. Therefore, the unknown data point must be the fourth or fifth data point, meaning it must be greater than or equal to 80. The only answer that fails to meet this condition is 60.

44. B: If 60% of 50 workers are women, then there are 30 women working in the office. If half of them are wearing skirts, then that means 15 women wear skirts. Since none of the men wear skirts, this means there are 15 people wearing skirts.

45. A: Let the unknown score be *x*. The average will be $\frac{5\times50+4\times70+x}{10} = \frac{530+x}{10} = 55$. Multiply both sides by 10 to get $530 + x = 550$, or $x = 20$.

46. D: For manufacturing costs, there is a linear relationship between the cost to the company and the number produced, with a *y*-intercept given by the base cost of acquiring the means of production, and a slope given by the cost to produce one unit. In this case, that base cost is \$50,000, while the cost per unit is \$40. So, $y = 40x + 50{,}000$.

47. C: A die has an equal chance for each outcome. Since it has six sides, each outcome has a probability of $\frac{1}{6}$. The chance of a 1 or a 2 is therefore $\frac{1}{6}+\frac{1}{6}=\frac{1}{3}$.

48. A: The slope is given by $m=\frac{y_2-y_1}{x_2-x_1}=\frac{0-4}{0-(-3)}=-\frac{4}{3}$.

49. B: An equilateral triangle has three sides of equal length, so if the total perimeter is 18 feet, each side must be 6 feet long. A square with sides of 6 feet will have an area of $6^2=36$ square feet.

50. B: The determinant of a 2 x 2 matrix is $ad-bc$. The calculation is -4(-1) – 2(3) = 4 – 6 = -2.

51. B: $\frac{2}{a}$ must be subtracted from both sides, with a result of $\frac{1}{x}=\frac{2}{b}-\frac{2}{a}$. The reciprocal of both sides needs to be taken, but the right-hand side needs to be written as a single fraction in order to do that. Since the two fractions on the right have denominators that are not equal, a common denominator of *ab* is needed. This leaves $\frac{1}{x}=\frac{2a}{ab}-\frac{2b}{ab}=\frac{2(a-b)}{ab}$. Taking the reciprocals, which can be done since *b* – *a* is not zero, with a result of $x=\frac{ab}{2(a-b)}$.

52. B: The first step is to use the quadratic formula on the first equation ($x^2+x-3=0$) to solve for *x*. In this case, *a* is 1, *b* is 1, and *c* is -3, yielding:

$$x=\frac{-b\pm\sqrt{b^2-4ac}}{2a}$$

$$x=\frac{-1\pm\sqrt{1-4\times 1(-3)}}{2}$$

$$x=\frac{-1}{2}\pm\frac{\sqrt{13}}{2}$$

Therefore, $x+\frac{1}{2}$, which is in our second equation, equals $\pm\frac{\sqrt{13}}{2}$. We are looking for $\left(x+\frac{1}{2}\right)^2$ though, so we square the $\pm\frac{\sqrt{13}}{2}$. Doing so causes the $\pm$ cancels and left with $\left(\frac{\sqrt{13}}{2}\right)^2=\frac{13}{4}$

53. A: Parallel lines have the same slope. The slope of *C* can be seen to be 1/3 by dividing both sides by 3. The others are in standard form $Ax+By=C$, for which the slope is given by $\frac{-A}{B}$. The slope of A is 3; the slope of B is 4. The slope of *D* is 1.

54. B: The slope will be given by $\frac{1-0}{2-0}=\frac{1}{2}$. The *y*-intercept will be 0, since it passes through the origin. Using slope-intercept form, the equation for this line is $y=\frac{1}{2}x$.

55. C: We are trying to find x, the number of red cans. The equation can be set up like this:

$$x+2(10-x)=16$$

The left x is actually multiplied by \$1, the price per red can. Since we know Jessica bought 10 total cans, $10-x$ is the number blue cans that she bought. We multiply the number of blue cans by \$2, the price per blue can.

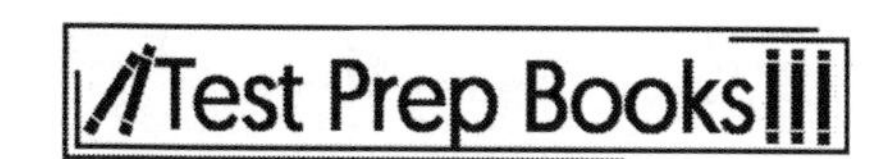

That should all equal $16, the total amount of money that Jessica spent. Working that out gives us:

$$x + 20 - 2x = 16$$

$$20 - x = 16$$

$$x = 4$$

56. D: The first field has an area of 1000 feet, so the length of one side is $\sqrt{1000} = 10\sqrt{10}$. Since there are four sides to a square, the total perimeter is $40\sqrt{10}$. The second square has an area of 10 square feet, so the length of one side is $\sqrt{10}$, and the total perimeter is $4\sqrt{10}$. Adding these together gives $40\sqrt{10} + 4\sqrt{10} = (40 + 4)\sqrt{10} = 44\sqrt{10}$.

57. C: If $\log_{10} x = 2$, then $10^2 = x$, which equals 100.

58. B: Recall that to compose functions, replace the *x* in the expression for *g* with the expression for *f* everywhere there is x. So:

$$g(f(x)) = \frac{f(x) - 1}{4} = \frac{2x + 1 - 1}{4} = \frac{2x}{4} = \frac{x}{2}$$

59. A: For acute angles, the only angle for which $\sin\theta = \frac{\sqrt{3}}{2}$ is $\frac{\pi}{3}$. Also, $\cos\frac{\pi}{3} = \frac{1}{2}$.

60. C: The $(3x + 1)$ can be factored to get $(2x - 5)(3x + 1)$.

61. D: Because the coefficient of x^2 is negative, this function has a graph that is a parabola that opens downward. Therefore, it will be greater than 0 between its real roots, if it has any. Checking the discriminant, the result is $1^2 - 4(-3)(-8) = 1 - 96 = -95$. Since the discriminant is negative, this equation has no real solutions. Since this has no real roots, it must be always positive or always negative. Its graph opens downward, so it has at least some negative values. That means it is always negative. Thus, it is greater than zero for no real numbers.

62. A: Check each value, but it is easiest to use the quadratic formula, which gives:

$$x = \frac{2 \pm \sqrt{(-2)^2 - 4(1)(-2)}}{2} = 1 \pm \frac{\sqrt{12}}{2} = 1 \pm \frac{2\sqrt{3}}{2} = 1 \pm \sqrt{3}$$

The only one of these which appears as an answer choice is $1 + \sqrt{3}$.

63. B: The *y* coordinate of every point on the graph of $y = x^2 + 2$ has a vertex at (0,2) on the y-axis. The circle with a center at (0,1) also lies on the y-axis. With a radius of 1, the circle touches the parabola at one point: the vertex of the parabola (0,2).

64. D: The slope is given by the change in *y* divided by the change in *x*. The change in *y* is 2-0 = 2, and the change in *x* is 0 – (-4) = 4. The slope is $\frac{2}{4} = \frac{1}{2}$.

65. B: The number of ways to order *n* objects is given by the product $\binom{n}{1}\binom{n-1}{1} \dots \binom{1}{1}$. This is $\binom{3}{1}\binom{2}{1}\binom{1}{1} = \frac{3!}{1!2!} \times \frac{2!}{1!1!} \times \frac{1!}{1!1!}$. $1!$ is just 1, and the $2!$ in the numerator and denominator will cancel one another, with a result of $3! = 3 \times 2 \times 1 = 6$.

66. C: If $y = f^{-1}(6)$ then *y* must satisfy $f(y) = 6$. Substituting and solving for y yields $4y + 2 = 6$, then $4y = 4$, and $y = 1$.

67. C: Find $a_2 = 3a_1 - 1 = 3 \times 1 - 1 = 2$. Next, find $a_3 = 3a_2 - 1 = 3 \times 2 - 1 = 5$.

68. C: Janice will be choosing 4 employees out of a set of 6 applicants, so this will be given by the choice function. The following equation shows the choice function worked out:

$$\binom{6}{4} = \frac{6!}{4!\,(6-4)!} = \frac{6!}{4!\,(2)!} = \frac{6 \times 5 \times 4 \times 3 \times 2 \times 1}{4 \times 3 \times 2 \times 1 \times 2 \times 1} = \frac{6 \times 5}{2} = 15$$

69. B: Here, *f* is an exponential function whose base is less than 1. In this function, *f* is always decreasing. This means that when *a* is less than *b*, $f(a) > f(b)$.

Reading

Literary Analysis

The Purpose of a Passage

When it comes to an author's writing, readers should always identify a position or stance. No matter how objective a text may seem, readers should assume the author has preconceived beliefs. One can reduce the likelihood of accepting an invalid argument by looking for multiple articles on the topic, including those with varying opinions. If several opinions point in the same direction and are backed by reputable peer-reviewed sources, it's more likely the author has a valid argument. Positions that run contrary to widely held beliefs and existing data should invite scrutiny. There are exceptions to the rule, so be a careful consumer of information.

Though themes, symbols, and motifs are buried deep within the text and can sometimes be difficult to infer, an author's purpose is usually obvious from the beginning. No matter the genre or format, all authors are writing to persuade, inform, entertain, or express feelings. Often, these purposes are blended, with one dominating the rest. It's useful to learn to recognize the author's intent.

Persuasive writing is used to persuade or convince readers of something. It often contains two elements: the argument and the counterargument. The argument takes a stance on an issue, while the counterargument pokes holes in the opposition's stance. Authors rely on logic, emotion, and writer credibility to persuade readers to agree with them. If readers are opposed to the stance before reading, they are unlikely to adopt that stance. However, those who are undecided or committed to the same stance are more likely to agree with the author.

Informative writing tries to teach or inform. Workplace manuals, instructor lessons, statistical reports and cookbooks are examples of informative texts. Informative writing is usually based on facts and is often void of emotion and persuasion. Informative texts generally contain statistics, charts, and graphs. Though most informative texts lack a persuasive agenda, readers must examine the text carefully to determine whether one exists within a given passage.

Stories or narratives are designed to entertain. When you go to the movies, you often want to escape for a few hours, not necessarily to think critically. Entertaining writing is designed to delight and engage the reader. However, sometimes this type of writing can be woven into more serious materials, such as persuasive or informative writing to hook the reader before transitioning into a more scholarly discussion.

Emotional writing works to evoke the reader's feelings, such as anger, euphoria, or sadness. The connection between reader and author is an attempt to cause the reader to share the author's intended emotion or tone. Sometimes in order to make a piece more poignant, the author simply wants readers to feel the same emotions that the author has felt. Other times, the author attempts to persuade or manipulate the reader into adopting his stance. While it's okay to sympathize with the author, be aware of the individual's underlying intent.

The various writing styles are usually blended, with one purpose dominating the rest. A persuasive text, for example, might begin with a humorous tale to make readers more receptive to the persuasive message, or a recipe in a cookbook designed to inform might be preceded by an entertaining anecdote that makes the recipes more appealing.

Identify Passage Characteristics

Writing can be classified under four passage types: narrative, expository, descriptive (sometimes called technical), and persuasive. Though these types are not mutually exclusive, one form tends to dominate the rest. By recognizing the *type* of passage you're reading, you gain insight into *how* you should read. When reading a narrative intended to entertain, sometimes you can read more quickly through the passage if the details are discernible. A technical document, on the other hand, might require a close read, because skimming the passage might cause the reader to miss salient details.

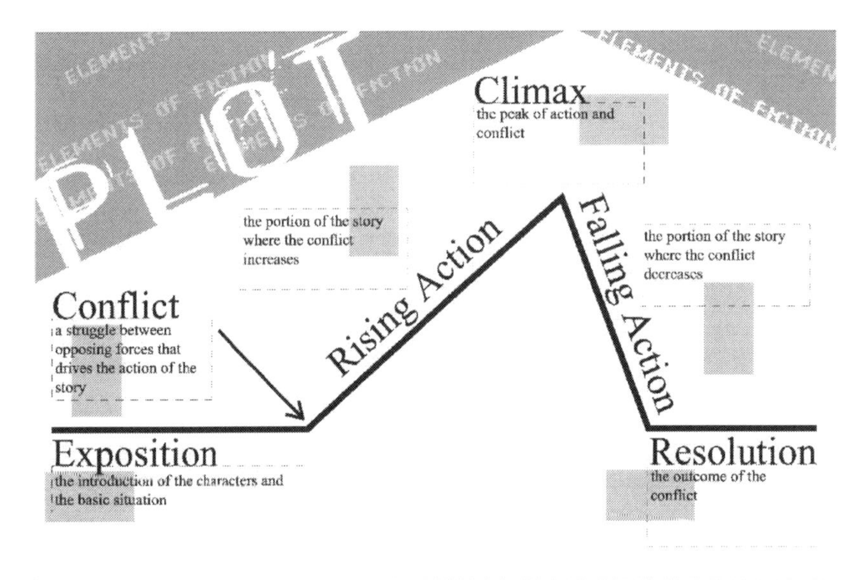

1. Narrative writing, at its core, is the art of storytelling. For a narrative to exist, certain elements must be present. First, it must have characters While many characters are human, characters could be defined as anything that thinks, acts, and talks like a human. For example, many recent movies, such as *Lord of the Rings* and *The Chronicles of Narnia*, include animals, fantasy creatures, and even trees that behave like humans. Narratives also must have a plot or sequence of events. Typically, those events follow a standard plot diagram, but recent trends start *in medias res* or in the middle (nearer the climax). In this instance, foreshadowing and flashbacks often fill in plot details. Finally, along with characters and a plot, there must also be conflict. Conflict is usually divided into two types: internal and external. Internal conflict indicates the character is in turmoil. Think of an angel on one shoulder and the devil on the other, arguing it out. Internal conflicts are presented through the character's thoughts. External conflicts are visible. Types of external conflict include person versus person, person versus nature, person versus technology, person versus the supernatural, or a person versus fate.

2. Expository writing is detached and to the point, while other types of writing — persuasive, narrative, and descriptive — are livelier. Since expository writing is designed to instruct or inform, it usually involves directions and steps written in second person ("you" voice) and lacks any persuasive or narrative elements. Sequence words such as *first*, *second*, and *third*, or *in the first place*, *secondly*, and *lastly* are often given to add fluency and cohesion. Common examples of expository writing include instructor's lessons, cookbook recipes, and repair manuals.

3. Due to its empirical nature, technical writing is filled with steps, charts, graphs, data, and statistics. The goal of technical writing is to advance understanding in a field through the scientific method. Experts such as teachers, doctors, or mechanics use words unique to the profession in which they operate. These words, which often incorporate acronyms, are called *jargon*. Technical writing is a type of expository writing but is not meant to be understood by the general public. Instead, technical writers assume readers have received a formal education in a particular field of study and need no explanation as to what the jargon means. Imagine a doctor trying to understand a diagnostic reading for a car or a mechanic trying to interpret lab results. Only professionals with proper training will fully comprehend the text.

4. Persuasive writing is designed to change opinions and attitudes. The topic, stance, and arguments are found in the thesis, positioned near the end of the introduction. Later supporting paragraphs offer relevant quotations, paraphrases, and summaries from primary or secondary sources, which are then interpreted, analyzed, and evaluated. The goal of persuasive writers is not to stack quotes, but to develop original ideas by using sources as a starting point. Good persuasive writing makes powerful arguments with valid sources and thoughtful analysis. Poor persuasive writing is riddled with bias and logical fallacies. Sometimes logical and illogical arguments are sandwiched together in the same text. Therefore, readers should display skepticism when reading persuasive arguments.

Text Structure

Depending on what the author is attempting to accomplish, certain formats or text structures work better than others. For instance, a sequence structure might work for narration but not when identifying similarities and differences between dissimilar concepts. Similarly, a comparison-contrast structure is not useful for narration. It's the author's job to put the right information in the correct format.

Readers should be familiar with the five main literary structures:

1. *Sequence* structure (sometimes referred to as the order structure) is when the order of events proceed in a predictable order. In many cases, this means the text goes through the plot elements: exposition, rising action, climax, falling action, and resolution. Readers are introduced to characters, setting, and conflict in the exposition. In the rising action, there's an increase in tension and suspense. The climax is the height of tension and the point of no return. Tension decreases during the falling action. In the resolution, any conflicts presented in the exposition are solved, and the story concludes. An informative text that is structured sequentially will often go in order from one step to the next.

2. In the *problem-solution* structure, authors identify a potential problem and suggest a solution. This form of writing is usually divided into two paragraphs and can be found in informational texts. For example, cell phone, cable, and satellite providers use this structure in manuals to help customers troubleshoot or identify problems with services or products.

3. When authors want to discuss similarities and differences between separate concepts, they arrange thoughts in a *comparison-contrast* paragraph structure. Venn diagrams are an effective graphic organizer for comparison-contrast structures because they feature two overlapping circles that can be used to organize similarities and differences. A comparison-contrast essay organizes one paragraph based on similarities and another based on differences. A comparison-contrast essay can also be arranged with the similarities and differences of individual traits addressed within individual paragraphs. Words such as *however*, *but*, and *nevertheless* help signal a contrast in ideas.

4. *Descriptive* writing structure is designed to appeal to your senses. Much like an artist who constructs a painting, good descriptive writing builds an image in the reader's mind by appealing to the five senses: sight, hearing, taste, touch, and smell. However, overly descriptive writing can become tedious; sparse descriptions can make settings and characters seem flat. Good authors strike a balance by applying descriptions only to passages, characters, and settings that are integral to the plot.

5. Passages that use the *cause and effect* structure are simply asking *why* by demonstrating some type of connection between ideas. Words such as *if*, *since*, *because*, *then*, or *consequently* indicate relationship. By switching the order of a complex sentence, the writer can rearrange the emphasis on different clauses. Saying *If Sheryl is late, we'll miss the dance* is different from saying *We'll miss the dance if Sheryl is late*. One emphasizes Sheryl's tardiness while the other emphasizes missing the dance. Paragraphs can also be arranged in a cause and effect format. Since the format—before and after—is sequential, it is useful when authors wish to discuss the impact of choices. Researchers often apply this paragraph structure to the scientific method.

Point of View

Point of view is an important writing device to consider. In fiction writing, point of view refers to who tells the story or from whose perspective readers are observing as they read. In non-fiction writing, the *point of view* refers to whether the author refers to himself/herself, his/her readers, or chooses not to mention either. Whether fiction or nonfiction, the author will carefully consider the impact the perspective will have on the purpose and main point of the writing.

- *First-person point of view*: The story is told from the writer's perspective. In fiction, this would mean that the main character is also the narrator. First-person point of view is easily recognized by the use of personal pronouns such as *I*, *me*, *we*, *us*, *our*, *my*, and *myself*.
- *Third-person point of view*: In a more formal essay, this would be an appropriate perspective because the focus should be on the subject matter, not the writer or the reader. Third-person point of view is recognized by the use of the pronouns *he*, *she*, *they*, and *it*. In fiction writing, third person point of view has a few variations.
    - *Third-person limited* point of view refers to a story told by a narrator who has access to the thoughts and feelings of just one character.
    - In *third-person omniscient* point of view, the narrator has access to the thoughts and feelings of all the characters.
    - In *third-person objective* point of view, the narrator is like a fly on the wall and can see and hear what the characters do and say but does not have access to their thoughts and feelings.

- *Second-person point of view*: This point of view isn't commonly used in fiction or non-fiction writing because it directly addresses the reader using the pronouns *you*, *your*, and *yourself*. Second-person perspective is more appropriate in direct communication, such as business letters or emails.

Point of View	Pronouns Used
First person	I, me, we, us, our, my, myself
Second person	You, your, yourself
Third person	He, she, it, they

Style, Tone, and Mood

Style, tone, and mood are often thought to be the same thing. Though they're closely related, there are important differences to keep in mind. The easiest way to do this is to remember that style "creates and affects" tone and mood. More specifically, style is how the writer uses words to create the desired tone and mood for their writing.

Style

Style can include any number of technical writing choices. A few examples of style choices include:

- Sentence Construction: When presenting facts, does the writer use shorter sentences to create a quicker sense of the supporting evidence, or do they use longer sentences to elaborate and explain the information?

- Technical Language: Does the writer use jargon to demonstrate their expertise in the subject, or do they use ordinary language to help the reader understand things in simple terms?

- Formal Language: Does the writer refrain from using contractions such as *won't* or *can't* to create a more formal tone, or do they use a colloquial, conversational style to connect to the reader?

- Formatting: Does the writer use a series of shorter paragraphs to help the reader follow a line of argument, or do they use longer paragraphs to examine an issue in great detail and demonstrate their knowledge of the topic?

On the test, examine the writer's style and how their writing choices affect the way the text comes across.

Tone

Tone refers to the writer's attitude toward the subject matter. Tone is usually explained in terms of a work of fiction. For example, the tone conveys how the writer feels about their characters and the situations in which they're involved. Nonfiction writing is sometimes thought to have no tone at all; however, this is incorrect.

A lot of nonfiction writing has a neutral tone, which is an important tone for the writer to take. A neutral tone demonstrates that the writer is presenting a topic impartially and letting the information speak for itself. On the other hand, nonfiction writing can be just as effective and appropriate if the tone isn't neutral. For instance, take this example involving seat belts:

> Seat belts save more lives than any other automobile safety feature. Many studies show that airbags save lives as well; however, not all cars have airbags. For instance, some older cars don't.

> Furthermore, air bags aren't entirely reliable. For example, studies show that in 15% of accidents airbags don't deploy as designed, but, on the other hand, seat belt malfunctions are extremely rare. The number of highway fatalities has plummeted since laws requiring seat belt usage were enacted.

In this passage, the writer mostly chooses to retain a neutral tone when presenting information. If the writer would instead include their own personal experience of losing a friend or family member in a car accident, the tone would change dramatically. The tone would no longer be neutral and would show that the writer has a personal stake in the content, allowing them to interpret the information in a different way. When analyzing tone, consider what the writer is trying to achieve in the text and how they *create* the tone using style.

Mood

Mood refers to the feelings and atmosphere that the writer's words create for the reader. Like tone, many nonfiction texts can have a neutral mood. To return to the previous example, if the writer would choose to include information about a person they know being killed in a car accident, the text would suddenly carry an emotional component that is absent in the previous example. Depending on how they present the information, the writer can create a sad, angry, or even hopeful mood. When analyzing the mood, consider what the writer wants to accomplish and whether the best choice was made to achieve that end.

Consistency

Whatever style, tone, and mood the writer uses, good writing should remain consistent throughout. If the writer chooses to include the tragic, personal experience above, it would affect the style, tone, and mood of the entire text. It would seem out of place for such an example to be used in the middle of a neutral, measured, and analytical text. To adjust the rest of the text, the writer needs to make additional choices to remain consistent. For example, the writer might decide to use the word *tragedy* in place of the more neutral *fatality*, or they could describe a series of car-related deaths as an *epidemic*. Adverbs and adjectives such as *devastating* or *horribly* could be included to maintain this consistent attitude toward the content. When analyzing writing, look for sudden shifts in style, tone, and mood, and consider whether the writer would be wiser to maintain the prevailing strategy.

Interpret Influences of Historical Context

Studying historical literature is fascinating. It reveals a snapshot in time of people, places, and cultures; a collective set of beliefs and attitudes that no longer exist. Writing changes as attitudes and cultures evolve. Beliefs previously considered immoral or wrong may be considered acceptable today. Researching the historical period of an author gives the reader perspective. The dialogue in Jane Austen's *Pride and Prejudice*, for example, is indicative of social class during the Regency era. Similarly, the stereotypes and slurs in *The Adventures of Huckleberry Finn* were a result of common attitudes and beliefs in the late 1800s, attitudes now found to be reprehensible.

Recognizing Cultural Themes

Regardless of culture, place, or time, certain themes are universal to the human condition. Because humans experience joy, rage, jealousy, and pride, certain themes span centuries. For example, Shakespeare's *Macbeth,* as well as modern works like *The 50th Law* by rapper 50 Cent and Robert Greene or the Netflix series *House of Cards* all feature characters who commit atrocious acts because of ambition. Similarly, *The Adventures of Huckleberry Finn*, published in the 1880s, and *The Catcher in the Rye*, published in the 1950s, both have characters who lie, connive, and survive on their wits.

Moviegoers know whether they are seeing an action, romance or horror film, and are often disappointed if the movie doesn't fit into the conventions of a particular category. Similarly, categories or genres give readers a sense of what to expect from a text. Some of the most basic genres in literature include books, short stories, poetry, and drama. Many genres can be split into sub-genres. For example, the sub-genres of historical fiction, realistic fiction, and fantasy all fit under the fiction genre.

Each genre has a unique way of approaching a particular theme. Books and short stories use plot, characterization, and setting, while poems rely on figurative language, sound devices, and symbolism. Dramas reveal plot through dialogue and the actor's voice and body language.

Main Ideas and Supporting Details

Topic Versus Main Idea

It is very important to know the difference between the topic and the main idea of the text. Even though these two are similar because they both present the central point of a text, they have distinctive differences. A *topic* is the subject of the text; it can usually be described in a one- to two-word phrase and appears in the simplest form. On the other hand, the *main idea* is more detailed and provides the author's central point of the text. It can be expressed through a complete sentence and can be found in the beginning, middle, or end of a paragraph. In most nonfiction books, the first sentence of the passage usually (but not always) states the main idea. Take a look at the passage below to review the topic versus the main idea.

Cheetahs

> Cheetahs are one of the fastest mammals on land, reaching up to 70 miles an hour over short distances. Even though cheetahs can run as fast as 70 miles an hour, they usually only have to run half that speed to catch up with their choice of prey. Cheetahs cannot maintain a fast pace over long periods of time because they will overheat their bodies. After a chase, cheetahs need to rest for approximately 30 minutes prior to eating or returning to any other activity.

In the example above, the topic of the passage is "Cheetahs" simply because that is the subject of the text. The main idea of the text is "Cheetahs are one of the fastest mammals on land but can only maintain this fast pace for short distances." While it covers the topic, it is more detailed and refers to the text in its entirety. The text continues to provide additional details called *supporting details,* which will be discussed in the next section.

Supporting Details

Supporting details help readers better develop and understand the main idea. Supporting details answer questions like *who, what, where, when, why,* and *how*. Different types of supporting details include examples, facts and statistics, anecdotes, and sensory details.

Persuasive and informative texts often use supporting details. In persuasive texts, authors attempt to make readers agree with their point of view, and supporting details are often used as "selling points." If authors make a statement, they should support the statement with evidence in order to adequately persuade readers. Informative texts use supporting details such as examples and facts to inform readers.

Take another look at the previous "Cheetahs" passage to find examples of supporting details.

Cheetahs

> Cheetahs are one of the fastest mammals on land, reaching up to 70 miles an hour over short distances. Even though cheetahs can run as fast as 70 miles an hour, they usually only have to run half that speed to catch up with their choice of prey. Cheetahs cannot maintain a fast pace over long periods of time because they will overheat their bodies. After a chase, cheetahs need to rest for approximately 30 minutes prior to eating or returning to any other activity.

In the example above, supporting details include:

- Cheetahs reach up to 70 miles per hour over short distances.
- They usually only have to run half that speed to catch up with their prey.
- Cheetahs will overheat their bodies if they exert a high speed over longer distances.
- Cheetahs need to rest for 30 minutes after a chase.

Look at the diagram below (applying the cheetah example) to help determine the hierarchy of topic, main idea, and supporting details.

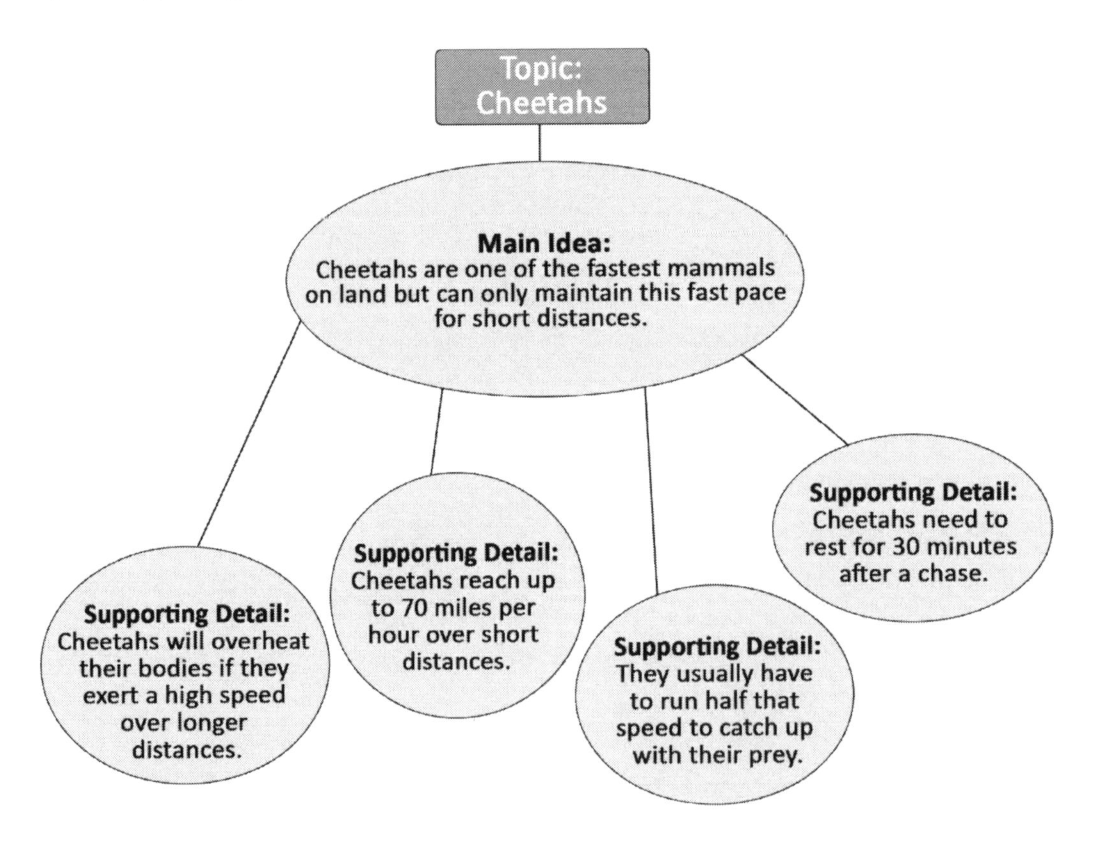

Drawing Conclusions

Determining conclusions requires being an active reader, as a reader must make a prediction and analyze facts to identify a conclusion. There are a few ways to determine a logical conclusion, but careful reading is the most important. It's helpful to read a passage a few times, noting details that seem important to the piece. A reader should also identify key words in a passage to determine the logical conclusion or determination that flows from the information presented.

Textual evidence within the details helps readers draw a conclusion about a passage. *Textual evidence* refers to information—facts and examples that support the main point. Textual evidence will likely come from outside sources and can be in the form of quoted or paraphrased material. In order to draw a conclusion from evidence, it's important to examine the credibility and validity of that evidence as well as how (and if) it relates to the main idea.

If an author presents a differing opinion or a *counter-argument* in order to refute it, the reader should consider how and why this information is being presented. It is meant to strengthen the original argument and shouldn't be confused with the author's intended conclusion, but it should also be considered in the reader's final evaluation.

Sometimes, authors explicitly state the conclusion they want readers to understand. Alternatively, a conclusion may not be directly stated. In that case, readers must rely on the implications to form a logical conclusion:

> On the way to the bus stop, Michael realized his homework wasn't in his backpack. He ran back to the house to get it and made it back to the bus just in time.

In this example, though it's never explicitly stated, it can be inferred that Michael is a student on his way to school in the morning. When forming a conclusion from implied information, it's important to read the text carefully to find several pieces of evidence in the text to support the conclusion.

Summarizing is an effective way to draw a conclusion from a passage. A summary is a shortened version of the original text, written by the reader in his/her own words. Focusing on the main points of the original text and including only the relevant details can help readers reach a conclusion. It's important to retain the original meaning of the passage.

Like summarizing, *paraphrasing* can also help a reader fully understand different parts of a text. Paraphrasing calls for the reader to take a small part of the passage and list or describe its main points. Paraphrasing is more than rewording the original passage, though. It should be written in the reader's own words, while still retaining the meaning of the original source. This will indicate an understanding of the original source, yet still help the reader expand on his/her interpretation.

Readers should pay attention to the *sequence*, or the order in which details are laid out in the text, as this can be important to understanding its meaning as a whole. Writers will often use transitional words to help the reader understand the order of events and to stay on track. Words like *next, then, after*, and *finally* show that the order of events is important to the author. In some cases, the author omits these transitional words, and the sequence is implied. Authors may even purposely present the information out of order to make an impact or have an effect on the reader. An example might be when a narrative writer uses *flashback* to reveal information.

There are several ways readers can draw conclusions from authors' ideas, such as note taking, text evidence, text credibility, writing a response to text, directly stated information versus implications, outlining, summarizing, and paraphrasing. Let's take a look at each important strategy to help readers draw logical conclusions.

Note Taking

When readers take notes throughout texts or passages, they are jotting down important facts or points that the author makes. Note taking is a useful record of information that helps readers understand the text or passage and respond to it. When taking notes, readers should keep lines brief and filled with pertinent information so that they are not rereading a large amount of text, but rather just key points, elements, or words. After readers have completed a text or passage, they can refer to their notes to help them form a conclusion about the author's ideas in the text or passage.

Text Evidence

Text evidence is the information readers find in a text or passage that supports the main idea or point(s) in a story. In turn, text evidence can help readers draw conclusions about the text or passage. The information should be taken directly from the text or passage and placed in quotation marks. Text evidence provides readers with information to support ideas about the text so that they do not rely simply on their own thoughts. Details should be precise, descriptive, and factual. Statistics are a great piece of text evidence because they provide readers with exact numbers and not just a generalization. For example, instead of saying "Asia has a larger population than Europe," authors could provide detailed information such as, "In Asia there are over 4 billion people, whereas in Europe there are a little over 750 million." More definitive information provides better evidence to readers to help support their conclusions about texts or passages.

Text Credibility

Credible sources are important when drawing conclusions because readers need to be able to trust what they are reading. Authors should always use credible sources to help gain the trust of their readers. A text is *credible* when it is believable and the author is objective and unbiased. If readers do not trust an author's words, they may simply dismiss the text completely. For example, if an author writes a persuasive essay, he or she is outwardly trying to sway readers' opinions to align with his or her own. Readers may agree or disagree with the author, which may, in turn, lead them to believe that the author is credible or not credible. Also, readers should keep in mind the source of the text. If readers review a journal about astronomy, would a more reliable source be a NASA employee or a medical doctor? Overall, text credibility is important when drawing conclusions, because readers want reliable sources that support the decisions they have made about the author's ideas.

Writing a Response to Text

Once readers have determined their opinions and validated the credibility of a text, they can then reflect on the text. Writing a response to a text is one method readers can reflect on the given text or passage. When readers write responses to a text, it is important for them to rely on the evidence within the text to support their opinions or thoughts. Supporting evidence such as facts, details, statistics, and quotes directly from the text are key pieces of information readers should reflect upon or use when writing a response to text.

Directly Stated Information Versus Implications

Engaged readers should constantly self-question while reviewing texts to help them form conclusions. Self-questioning is when readers review a paragraph, page, passage, or chapter and ask themselves, "Did I understand what I read?", "What was the main event in this section?", "Where is this taking place?", and so on. Authors can provide clues or pieces of evidence throughout a text or passage to guide readers toward a conclusion. This is why active and engaged readers should read the text or passage in its entirety before forming a definitive conclusion. If readers do not gather all the pieces of evidence needed, then they may jump to an illogical conclusion.

At times, authors directly state conclusions while others simply imply them. Of course, it is easier if authors outwardly provide conclusions to readers because it does not leave any information open to interpretation. On the other hand, implications are things that authors do not directly state but can be assumed based off of information they provided. If authors only imply what may have happened, readers can form a menagerie of ideas for conclusions. For example, look at the following statement: "Once we heard the sirens, we hunkered down in the storm shelter." In this statement, the author does not directly state that there was a tornado, but clues such as "sirens" and "storm shelter" provide insight to the readers to help form that conclusion.

Outlining

An outline is a system used to organize writing. When reading texts, outlining is important because it helps readers organize important information in a logical pattern using roman numerals. Usually, outlines start with the main idea(s) and then branch out into subgroups or subsidiary thoughts of subjects. Not only do outlines provide a visual tool for readers to reflect on how events, characters, settings, or other key parts of the text or passage relate to one another, but they can also lead readers to a stronger conclusion.

The sample below demonstrates what a general outline looks like.

I. Main Topic 1
    a. Subtopic 1
    b. Subtopic 2
        1. Detail 1
        2. Detail 2
II. Main Topic 2
    a. Subtopic 1
    b. Subtopic 2
        1. Detail 1
        2. Detail 2

Summarizing

At the end of a text or passage, it is important to summarize what the readers read. Summarizing is a strategy in which readers determine what is important throughout the text or passage, shorten those ideas, and rewrite or retell it in their own words. A summary should identify the main idea of the text or passage. Important details or supportive evidence should also be accurately reported in the summary. If writers provide irrelevant details in the summary, it may cloud the greater meaning of the summary in the text. When summarizing, writers should not include their opinions, quotes, or what they thought the author should have said. A clear summary provides clarity of the text or passage to the readers. Let's review the checklist of items writers should include in their summary.

Summary Checklist

- Title of the story
- Someone: Who is or are the main character(s)?
- Wanted: What did the character(s) want?
- But: What was the problem?
- So: How did the character(s) solve the problem?
- Then: How did the story end? What was the resolution?

Paraphrasing

Another strategy readers can use to help them fully comprehend a text or passage is paraphrasing. Paraphrasing is when readers take the author's words and put them into their own words. When readers and writers paraphrase, they should avoid copying the text—that is plagiarism. It is also important to include as many details as possible when restating the facts. Not only will this help readers and writers recall information, but by putting the information into their own words, they demonstrate whether or not they fully comprehend the text or passage. Look at the example below showing an original text and how to paraphrase it.

> *Original Text*: Fenway Park is home to the beloved Boston Red Sox. The stadium opened on April 20, 1912. The stadium currently seats over 37,000 fans, many of whom travel from all over the country to experience the iconic team and nostalgia of Fenway Park.
>
> *Paraphrased*: On April 20, 1912, Fenway Park opened. Home to the Boston Red Sox, the stadium now seats over 37,000 fans. Many spectators travel to watch the Red Sox and experience the spirit of Fenway Park.

Paraphrasing, summarizing, and quoting can often cross paths with one another. Review the chart below showing the similarities and differences between the three strategies.

Paraphrasing	Summarizing	Quoting
Uses own words	Puts main ideas into own words	Uses words that are identical to text
References original source	References original source	Requires quotation marks
Uses own sentences	Shows important ideas of source	Uses author's own words and ideas

Inferences in a Text

Readers should be able to make *inferences*. Making an inference requires the reader to read between the lines and look for what is *implied* rather than what is directly stated. That is, using information that is known from the text, the reader is able to make a logical assumption about information that is *not* directly stated but is probably true. Read the following passage:

"Hey, do you wanna meet my new puppy?" Jonathan asked.

"Oh, I'm sorry but please don't—" Jacinta began to protest, but before she could finish, Jonathan had already opened the passenger side door of his car and a perfect white ball of fur came bouncing towards Jacinta.

"Isn't he the cutest?" beamed Jonathan.

"Yes—achoo!—he's pretty—aaaachooo!!—adora—aaa—aaaachoo!" Jacinta managed to say in between sneezes. "But if you don't mind, I—I—achoo!—need to go inside."

Which of the following can be inferred from Jacinta's reaction to the puppy?

a. she hates animals
b. she is allergic to dogs
c. she prefers cats to dogs
d. she is angry at Jonathan

An inference requires the reader to consider the information presented and then form their own idea about what is probably true. Based on the details in the passage, what is the best answer to the question? Important details to pay attention to include the tone of Jacinta's dialogue, which is overall polite and apologetic, as well as her reaction itself, which is a long string of sneezes. Answer choices (a) and (d) both express strong emotions ("hates" and "angry") that are not evident in Jacinta's speech or actions. Answer choice (c) mentions cats, but there is nothing in the passage to indicate Jacinta's feelings about cats. Answer choice (b), "she is allergic to dogs," is the most logical choice—based on the fact that she began sneezing as soon as a fluffy dog approached her, it makes sense to guess that Jacinta might be allergic to dogs. So even though Jacinta never directly states, "Sorry, I'm allergic to dogs!" using the clues in the passage, it is still reasonable to guess that this is true.

Making inferences is crucial for readers of literature because literary texts often avoid presenting complete and direct information to readers about characters' thoughts or feelings, or they present this information in an unclear way, leaving it up to the reader to interpret clues given in the text. In order to make inferences while reading, readers should ask themselves:

- What details are being presented in the text?
- Is there any important information that seems to be missing?
- Based on the information that the author *does* include, what else is probably true?
- Is this inference reasonable based on what is already known?

Apply Information

A natural extension of being able to make an inference from a given set of information is also being able to apply that information to a new context. This is especially useful in non-fiction or informative writing. Considering the facts and details presented in the text, readers should consider how the same information might be relevant in a different situation. The following is an example of applying an inferential conclusion to a different context:

Often, individuals behave differently in large groups than they do as individuals. One example of this is the psychological phenomenon known as the bystander effect. According to the bystander effect, the more people who witness an accident or crime occur, the less likely each individual bystander is to respond or offer assistance to the victim. A classic example of this is the murder of Kitty Genovese in New York City in the 1960s. Although there were over thirty witnesses to her killing by a stabber, none of them intervened to help Kitty or contact the police.

Considering the phenomenon of the bystander effect, what would probably happen if somebody tripped on the stairs in a crowded subway station?

a. Everybody would stop to help the person who tripped
b. Bystanders would point and laugh at the person who tripped
c. Someone would call the police after walking away from the station
d. Few if any bystanders would offer assistance to the person who tripped

This question asks readers to apply the information they learned from the passage, which is an informative paragraph about the bystander effect. According to the passage, this is a concept in psychology that describes the way people in groups respond to an accident—the more people are present, the less likely any one person is to intervene. While the passage illustrates this effect with the example of a woman's murder, the question asks readers to apply it to a different context—in this case, someone falling down the stairs in front of many subway passengers. Although this specific situation is not discussed in the passage, readers should be able to apply the general concepts described in the paragraph. The definition of the bystander effect includes any instance of an accident or crime in front of a large group of people. The question asks about a situation that falls within the same definition, so the general concept should still hold true: in the midst of a large crowd, few individuals are likely to actually respond to an accident. In this case, answer choice (d) is the best response.

Critical Thinking Skills

It's important to read any piece of writing critically. The goal is to discover the point and purpose of what the author is writing about through analysis. It's also crucial to establish the point or stance the author has taken on the topic of the piece. After determining the author's perspective, readers can then more effectively develop their own viewpoints on the subject of the piece.

It is important to distinguish between *fact and opinion* when reading a piece of writing. A fact is information that can be proven true. If information can be disproved, it is not a fact. For example, water freezes at or below thirty-two degrees Fahrenheit. An argument stating that water freezes at seventy degrees Fahrenheit cannot be supported by data and is therefore not a fact. Facts tend to be associated with science, mathematics, and statistics. Opinions are information open to debate. Opinions are often tied to subjective concepts like equality, morals, and rights. They can also be controversial.

Authors often use words like *think, feel, believe,* or *in my opinion* when expressing opinion, but these words won't always appear in an opinion piece, especially if it is formally written. An author's opinion may be backed up by facts, which gives it more credibility, but that opinion should not be taken as fact. A critical reader should be suspect of an author's opinion, especially if it is only supported by other opinions.

Fact	Opinion
There are nine innings in a game of baseball.	Baseball games run too long.
James Garfield was assassinated on July 2, 1881.	James Garfield was a good president.
McDonalds has stores in 118 countries.	McDonalds has the best hamburgers.

Critical readers examine the facts used to support an author's argument. They check the facts against other sources to be sure those facts are correct. They also check the validity of the sources used to be sure those sources are credible, academic, and/or peer-reviewed. Consider that when an author uses another person's opinion to support his or her argument, even if it is an expert's opinion, it is still only an opinion and should not be taken as fact. A strong argument uses valid, measurable facts to support ideas. Even then, the reader may disagree with the argument as it may be rooted in his or her personal beliefs.

An authoritative argument may use the facts to sway the reader. In the example of global warming, many experts differ in their opinions of what alternative fuels can be used to aid in offsetting it. Because of this, a writer may choose to only use the information and expert opinion that supports his or her viewpoint.

If the argument is that wind energy is the best solution, the author will use facts that support this idea. That same author may leave out relevant facts on solar energy. The way the author uses facts can influence the reader, so it's important to consider the facts being used, how those facts are being presented, and what information might be left out.

Critical readers should also look for errors in the argument such as logical fallacies and bias. A *logical fallacy* is a flaw in the logic used to make the argument. Logical fallacies include slippery slope, straw man, and begging the question. Authors can also reflect *bias* if they ignore an opposing viewpoint or present their side in an unbalanced way. A strong argument considers the opposition and finds a way to refute it. Critical readers should look for an unfair or one-sided presentation of the argument and be skeptical, as a bias may be present. Even if this bias is unintentional, if it exists in the writing, the reader should be wary of the validity of the argument.

Readers should also look for the use of *stereotypes,* which refer to specific groups. Stereotypes are often negative connotations about a person or place and should always be avoided. When a critical reader finds stereotypes in a piece of writing, they should immediately be critical of the argument and consider the validity of anything the author presents. Stereotypes reveal a flaw in the writer's thinking and may suggest a lack of knowledge or understanding about the subject.

Author's Use of Language

Authors utilize a wide range of techniques to tell a story or communicate information. Readers should be familiar with the most common of these techniques. Techniques of writing are also commonly known as rhetorical devices.

Types of Appeals

In non-fiction writing, authors employ argumentative techniques to present their opinion to readers in the most convincing way. First of all, persuasive writing usually includes at least one type of appeal: an appeal to logic (logos), emotion (pathos), or credibility and trustworthiness (ethos). When a writer appeals to logic, they are asking readers to agree with them based on research, evidence, and an established line of reasoning. An author's argument might also appeal to readers' emotions, perhaps by including personal stories and anecdotes (a short narrative of a specific event). A final type of appeal, appeal to authority, asks the reader to agree with the author's argument on the basis of their expertise or credentials. Consider three different approaches to arguing the same opinion:

Logic (Logos)

This is an example of an appeal to logic:

> Our school should abolish its current ban on cell phone use on campus. This rule was adopted last year as an attempt to reduce class disruptions and help students focus more on their lessons. However, since the rule was enacted, there has been no change in the number of disciplinary problems in class. Therefore, the rule is ineffective and should be done away with.

The author uses evidence to disprove the logic of the school's rule (the rule was supposed to reduce discipline problems; the number of problems has not been reduced; therefore, the rule is not working) and call for its repeal.

Emotion (Pathos)

An author's argument might also appeal to readers' emotions, perhaps by including personal stories and anecdotes. The next example presents an appeal to emotion. By sharing the personal anecdote of one student and speaking about emotional topics like family relationships, the author invokes the reader's empathy in asking them to reconsider the school rule.

> Our school should abolish its current ban on cell phone use on campus. If they aren't able to use their phones during the school day, many students feel isolated from their loved ones. For example, last semester, one student's grandmother had a heart attack in the morning. However, because he couldn't use his cell phone, the student didn't know about his grandmother's accident until the end of the day—when she had already passed away, and it was too late to say goodbye. By preventing students from contacting their friends and family, our school is placing undue stress and anxiety on students.

Credibility (Ethos)

Finally, an appeal to authority includes a statement from a relevant expert. In this case, the author uses a doctor in the field of education to support the argument. All three examples begin from the same opinion—the school's phone ban needs to change—but rely on different argumentative styles to persuade the reader.

> Our school should abolish its current ban on cell phone use on campus. According to Dr. Bartholomew Everett, a leading educational expert, "Research studies show that cell phone usage has no real impact on student attentiveness. Rather, phones provide a valuable technological resource for learning. Schools need to learn how to integrate this new technology into their curriculum." Rather than banning phones altogether, our school should follow the advice of experts and allow students to use phones as part of their learning.

Rhetorical Questions

Another commonly used argumentative technique is asking rhetorical questions, questions that do not actually require an answer but that push the reader to consider the topic further.

> I wholly disagree with the proposal to ban restaurants from serving foods with high sugar and sodium contents. Do we really want to live in a world where the government can control what we eat? I prefer to make my own food choices.

Here, the author's rhetorical question prompts readers to put themselves in a hypothetical situation and imagine how they would feel about it.

Figurative Language

Literary texts also employ rhetorical devices. Figurative language like simile and metaphor is a type of rhetorical device commonly found in literature. In addition to rhetorical devices that play on the *meanings* of words, there are also rhetorical devices that use the *sounds* of words. These devices are most often found in poetry but may also be found in other types of literature and in non-fiction writing like speech texts.

Alliteration and *assonance* are both varieties of sound repetition. Other types of sound repetition include: anaphora, repetition that occurs at the beginning of the sentences; epiphora, repetition occurring at the end of phrases; antimetabole, repetition of words in reverse order; and antiphrasis, a form of denial of an assertion in a text.

Alliteration refers to the repetition of the first sound of each word. Recall Robert Burns' opening line:

> My love is like a red, red rose

This line includes two instances of alliteration: "love" and "like" (repeated *L* sound), as well as "red" and "rose" (repeated *R* sound). Next, assonance refers to the repetition of vowel sounds, and can occur anywhere within a word (not just the opening sound). Here is the opening of a poem by John Keats:

> When I have fears that I may cease to be
>
> Before my pen has glean'd my teeming brain

Assonance can be found in the words "fears," "cease," "be," "glean'd," and "teeming," all of which stress the long *E* sound. Both alliteration and assonance create a harmony that unifies the writer's language.

Another sound device is *onomatopoeia*, or words whose spelling mimics the sound they describe. Words such as "crash," "bang," and "sizzle" are all examples of onomatopoeia. Use of onomatopoetic language adds auditory imagery to the text.

Readers are probably most familiar with the technique of *pun*. A pun is a play on words, taking advantage of two words that have the same or similar pronunciation. Puns can be found throughout Shakespeare's plays, for instance:

> Now is the winter of our discontent
> Made glorious summer by this son of York

These lines from *Richard III* contain a play on words. Richard III refers to his brother, the newly crowned King Edward IV, as the "son of York," referencing their family heritage from the house of York. However, while drawing a comparison between the political climate and the weather (times of political trouble were the "winter," but now the new king brings "glorious summer"), Richard's use of the word "son" also implies another word with the same pronunciation, "sun"—so Edward IV is also like the sun, bringing light, warmth, and hope to England. Puns are a clever way for writers to suggest two meanings at once.

Counterarguments

If an author presents a differing opinion or a counterargument in order to refute it, the reader should consider how and why this information is being presented. It is meant to strengthen the original argument and shouldn't be confused with the author's intended conclusion, but it should also be considered in the reader's final evaluation.

Authors can also use bias if they ignore the opposing viewpoint or present their side in an unbalanced way. A strong argument considers the opposition and finds a way to refute it. Critical readers should look for an unfair or one-sided presentation of the argument and be skeptical, as a bias may be present. Even if this bias is unintentional, if it exists in the writing, the reader should be wary of the validity of the argument. Readers should also look for the use of stereotypes, which refer to specific groups. Stereotypes are often negative connotations about a person or place and should always be avoided. When a critical reader finds stereotypes in a piece of writing, they should be critical of the argument, and consider the validity of anything the author presents. Stereotypes reveal a flaw in the writer's thinking and may suggest a lack of knowledge or understanding about the subject.

Meaning of Words in Context

There will be many occasions in one's reading career in which an unknown word or a word with multiple meanings will pop up. There are ways of determining what these words or phrases mean that do not require the use of the dictionary, which is especially helpful during a test where one may not be available. Even outside of the exam, knowing how to derive an understanding of a word via context clues will be a critical skill in the real world. The context is the circumstances in which a story or a passage is happening and can usually be found in the series of words directly before or directly after the word or phrase in question. The clues are the words that hint towards the meaning of the unknown word or phrase.

There may be questions that ask about the meaning of a particular word or phrase within a passage. There are a couple ways to approach these kinds of questions:

1. Define the word or phrase in a way that is easy to comprehend (using context clues).
2. Try out each answer choice in place of the word.

To demonstrate, here's an example from *Alice in Wonderland*:

> Alice was beginning to get very tired of sitting by her sister on the bank, and of having nothing to do: once or twice she peeped into the book her sister was reading, but it had no pictures or conversations in it, "and what is the use of a book," thought Alice, "without pictures or conversations?"

Q: As it is used in the selection, the word peeped means:

Using the first technique, before looking at the answers, define the word "peeped" using context clues and then find the matching answer. Then, analyze the entire passage in order to determine the meaning, not just the surrounding words.

To begin, imagine a blank where the word should be and put a synonym or definition there: "once or twice she ______ into the book her sister was reading." The context clue here is the book. It may be tempting to put "read" where the blank is, but notice the preposition word, "into." One does not read *into* a book, one simply reads a book, and since reading a book requires that it is seen with a pair of eyes, then "look" would make the most sense to put into the blank: "once or twice she looked into the book her sister was reading."

Once an easy-to-understand word or synonym has been supplanted, readers should check to make sure it makes sense with the rest of the passage. What happened after she looked into the book? She thought to herself how a book without pictures or conversations is useless. This situation in its entirety makes sense.

Now check the answer choices for a match:

a. To make a high-pitched cry
b. To smack
c. To look curiously
d. To pout

Since the word was already defined, Choice *C* is the best option.

Using the second technique, replace the figurative blank with each of the answer choices and determine which one is the most appropriate. Remember to look further into the passage to clarify that they work, because they could still make sense out of context.

a. Once or twice she made a high-pitched cry into the book her sister was reading
b. Once or twice she smacked into the book her sister was reading
c. Once or twice she looked curiously into the book her sister was reading
d. Once or twice she pouted into the book her sister was reading

For Choice *A*, it does not make much sense in any context for a person to yell into a book, unless maybe something terrible has happened in the story. Given that afterward Alice thinks to herself how useless a book without pictures is, this option does not make sense within context.

For Choice *B*, smacking a book someone is reading may make sense if the rest of the passage indicates a reason for doing so. If Alice was angry or her sister had shoved it in her face, then maybe smacking the book would make sense within context. However, since whatever she does with the book causes her to think, "what is the use of a book without pictures or conversations?" then answer Choice *B* is not an appropriate answer. Answer Choice *C* fits well within context, given her subsequent thoughts on the matter. Answer Choice *D* does not make sense in context or grammatically, as people do not "pout into" things.

This is a simple example to illustrate the techniques outlined above. There may, however, be a question in which all of the definitions are correct and also make sense out of context, in which the appropriate context clues will really need to be honed in on in order to determine the correct answer. For example, here is another passage from *Alice in Wonderland*:

> . . . but when the Rabbit actually took a watch out of its waistcoat pocket, and looked at it, and then hurried on, Alice started to her feet, for it flashed across her mind that she had never before seen a rabbit with either a waistcoat-pocket or a watch to take out of it, and burning with curiosity, she ran across the field after it, and was just in time to see it pop down a large rabbit-hole under the hedge.

Q: As it is used in the passage, the word started means

a. To turn on
b. To begin
c. To move quickly
d. To be surprised

All of these words qualify as a definition of "start," but using context clues, the correct answer can be identified using one of the two techniques above. It's easy to see that one does not turn on, begin, or be surprised to one's feet. The selection also states that she "ran across the field after it," indicating that she was in a hurry. Therefore, to move quickly would make the most sense in this context.

The same strategies can be applied to vocabulary that may be completely unfamiliar. In this case, focus on the words before or after the unknown word in order to determine its definition. Take this sentence, for example:

> Sam was such a miser that he forced Andrew to pay him twelve cents for the candy, even though he had a large inheritance and he knew his friend was poor.

Unlike with assertion questions, for vocabulary questions, it may be necessary to apply some critical thinking skills that may not be explicitly stated within the passage. Think about the implications of the passage, or what the text is trying to say. With this example, it is important to realize that it is considered unusually stingy for a person to demand so little money from someone instead of just letting their friend have the candy, especially if this person is already wealthy. Hence, a miser is a greedy or stingy individual.

Questions about complex vocabulary may not be explicitly asked, but this is a useful skill to know. If there is an unfamiliar word while reading a passage and its definition goes unknown, it is possible to miss out on a critical message that could inhibit the ability to appropriately answer the questions. Practicing this technique in daily life will sharpen this ability to derive meanings from context clues with ease.

Practice Test

Directions for questions 1–9: Read the statement or passage and then choose the best answer to the question. Answer the question based on what is stated or implied in the statement or passage.

1. There are two major kinds of cameras on the market right now for amateur photographers. Camera enthusiasts can either purchase a digital single-lens reflex camera (DSLR) camera or a compact system camera (CSC). The main difference between a DSLR and a CSC is that the DSLR has a full-sized sensor, which means it fits in a much larger body. The CSC uses a mirrorless system, which makes for a lighter, smaller camera. While both take quality pictures, the DSLR generally has better picture quality due to the larger sensor. CSCs still take very good quality pictures and are more convenient to carry than a DSLR. This makes the CSC an ideal choice for the amateur photographer looking to step up from a point-and-shoot camera.

What is the main difference between the DSLR and CSC?

a. The picture quality is better in the DSLR.
b. The CSC is less expensive than the DSLR.
c. The DSLR is a better choice for amateur photographers.
d. The DSLR's larger sensor makes it a bigger camera than the CSC.

2. When selecting a career path, it's important to explore the various options available. Many students entering college may shy away from a major because they don't know much about it. For example, many students won't opt for a career as an actuary, because they aren't exactly sure what it entails. They would be missing out on a career that is very lucrative and in high demand. Actuaries work in the insurance field and assess risks and premiums. The average salary of an actuary is $100,000 per year. Another career option students may avoid, due to lack of knowledge of the field, is a hospitalist. This is a physician that specializes in the care of patients in a hospital, as opposed to those seen in private practices. The average salary of a hospitalist is upwards of $200,000. It pays to do some digging and find out more about these lesser-known career fields.

What is an actuary?

a. A doctor who works in a hospital.
b. The same as a hospitalist.
c. An insurance agent who works in a hospital.
d. A person who assesses insurance risks and premiums.

3. Hard water occurs when rainwater mixes with minerals from rock and soil. Hard water has a high mineral count, including calcium and magnesium. The mineral deposits from hard water can stain hard surfaces in bathrooms and kitchens as well as clog pipes. Hard water can stain dishes, ruin clothes, and reduce the life of any appliances it touches, such as hot water heaters, washing machines, and humidifiers.

One solution is to install a water softener to reduce the mineral content of water, but this can be costly. Running vinegar through pipes and appliances and using vinegar to clean hard surfaces can also help with mineral deposits.

From this passage, what can be concluded?

a. Hard water can cause a lot of problems for homeowners.
b. Calcium is good for pipes and hard surfaces.
c. Water softeners are easy to install.
d. Vinegar is the only solution to hard water problems.

4. Coaches of kids' sports teams are increasingly concerned about the behavior of parents at games. Parents are screaming and cursing at coaches, officials, players, and other parents. Physical fights have even broken out at games. Parents need to be reminded that coaches are volunteers, giving up their time and energy to help kids develop in their chosen sport. The goal of kids' sports teams is to learn and develop skills, but it's also to have fun. When parents are out of control at games and practices, it takes the fun out of the sport.

From this passage, what can be concluded?

a. Coaches are modeling good behavior for kids.
b. Organized sports are not good for kids.
c. Parents' behavior at their kids' games needs to change.
d. Parents and coaches need to work together.

5. While scientists aren't entirely certain why tornadoes form, they have some clues into the process. Tornadoes are dangerous funnel clouds that occur during a large thunderstorm. When warm, humid air near the ground meets cold, dry air from above, a column of the warm air can be drawn up into the clouds. Winds at different altitudes blowing at different speeds make the column of air rotate. As the spinning column of air picks up speed, a funnel cloud is formed. This funnel cloud moves rapidly and haphazardly. Rain and hail inside the cloud cause it to touch down, creating a tornado. Tornadoes move in a rapid and unpredictable pattern, making them extremely destructive and dangerous. Scientists continue to study tornadoes to improve radar detection and warning times.

The main purpose of this passage is to do which of the following?

a. Show why tornadoes are dangerous.
b. Explain how a tornado forms.
c. Compare thunderstorms to tornadoes.
d. Explain what to do in the event of a tornado.

6. Many people are unsure of exactly how the digestive system works. Digestion begins in the mouth where teeth grind up food, and saliva breaks it down, making it easier for the body to absorb. Next, the food moves to the esophagus, and it is pushed into the stomach. The stomach is where food is stored and broken down further by acids and digestive enzymes, preparing it for passage into the intestines. The small intestine is where the nutrients are taken from food and passed into the blood stream. Other essential organs like the liver, gall bladder, and pancreas aid the stomach in breaking down food and absorbing nutrients. Finally, food waste is passed into the large intestine where it is eliminated by the body.

The purpose of this passage is to do which of the following?

a. Explain how the liver works.
b. Show why it is important to eat healthy foods.
c. Explain how the digestive system works.
d. Show how nutrients are absorbed by the small intestine.

7. Osteoporosis is a medical condition that occurs when the body loses bone or makes too little bone. This can lead to brittle, fragile bones that easily break. Bones are already porous, and when osteoporosis sets in, the spaces in bones become much larger, causing them to weaken. Both men and women can contract osteoporosis, though it is most common in women over age 50. Loss of bone can be silent and progressive, so it is important to be proactive in prevention of the disease.

The main purpose of this passage is to do which of the following?

a. Discuss some of the ways people contract osteoporosis.
b. Describe different treatment options for those with osteoporosis.
c. Explain how to prevent osteoporosis.
d. Define osteoporosis.

8. Vacationers looking for a perfect experience should opt out of Disney parks and try a trip on Disney Cruise Lines. While a park offers rides, characters, and show experiences, it also includes long lines, often very hot weather, and enormous crowds. A Disney Cruise, on the other hand, is a relaxing, luxurious vacation that includes many of the same experiences as the parks, minus the crowds and lines. The cruise has top-notch food, maid service, water slides, multiple pools, Broadway-quality shows, and daily character experiences for kids. There are also many activities, such as bingo, trivia contests, and dance parties that can entertain guests of all ages. The cruise even stops at Disney's private island for a beach barbecue with characters, waterslides, and water sports. Those looking for the Disney experience without the hassle should book a Disney cruise.

The main purpose of this passage is to do which of the following?

a. Explain how to book a Disney cruise.
b. Show what Disney parks have to offer.
c. Show why Disney parks are expensive.
d. Compare Disney parks to a Disney cruise.

9. As summer approaches, drowning incidents will increase. Drowning happens very quickly and silently. Most people assume that drowning is easy to spot, but a person who is drowning doesn't make noise or wave their arms. Instead, they will have their head back and their mouth open, with just the face out of the water. A person who is truly in danger of drowning is not able to wave their arms in the air or move much at all. Recognizing these signs of drowning can prevent tragedy.

The main purpose of this passage is to do which of the following?
a. Explain the dangers of swimming.
b. Show how to identify the signs of drowning.
c. Explain how to be a lifeguard.
d. Compare the signs of drowning.

The next question is based on the following conversation between a scientist and a politician.

Scientist: Last year was the warmest ever recorded in the last 134 years. During that time period, the ten warmest years have all occurred since 2000. This correlates directly with the recent increases in carbon dioxide as large countries like China, India, and Brazil continue developing and industrializing. No longer do just a handful of countries burn massive amounts of carbon-based fossil fuels; it is quickly becoming the case throughout the whole world as technology and industry spread.

Politician: Yes, but there is no causal link between increases in carbon emissions and increasing temperatures. The link is tenuous and nothing close to certain. We need to wait for all of the data before drawing hasty conclusions. For all we know, the temperature increase could be entirely natural. I believe the temperatures also rose dramatically during the dinosaurs' time, and I do not think they were burning any fossil fuels back then.

10. What is one point on which the scientist and politician agree?
a. Burning fossil fuels causes global temperatures to rise.
b. Global temperatures are increasing.
c. Countries must revisit their energy policies before it's too late.
d. Earth's climate naturally goes through warming and cooling periods.

The next question is based on the following passage.

A famous children's author recently published a historical fiction novel under a pseudonym; however, it did not sell as many copies as her children's books. In her earlier years, she had majored in history and earned a graduate degree in Antebellum American History, which is the time frame of her new novel. Critics praised this newest work far more than the children's series that made her famous. In fact, her new novel was nominated for the prestigious Albert J. Beveridge Award but still isn't selling like her children's books, which fly off the shelves because of her name alone.

11. Which one of the following statements might be accurately inferred based on the above passage?
    a. The famous children's author produced an inferior book under her pseudonym.
    b. The famous children's author is the foremost expert on Antebellum America.
    c. The famous children's author did not receive the bump in publicity for her historical novel that it would have received if it were written under her given name.
    d. People generally prefer to read children's series than historical fiction.

The next three questions are based on the following passage.

Smoking is Terrible

> Smoking tobacco products is terribly destructive. A single cigarette contains over 4,000 chemicals, including 43 known carcinogens and 400 deadly toxins. Some of the most dangerous ingredients include tar, carbon monoxide, formaldehyde, ammonia, arsenic, and DDT. Smoking can cause numerous types of cancer including throat, mouth, nasal cavity, esophageal, gastric, pancreatic, renal, bladder, and cervical cancer.
>
> Cigarettes contain a drug called nicotine, one of the most addictive substances known to man. Addiction is defined as a compulsion to seek the substance despite negative consequences. According to the National Institute of Drug Abuse, nearly 35 million smokers expressed a desire to quit smoking in 2015; however, more than 85 percent of those who struggle with addiction will not achieve their goal. Almost all smokers regret picking up that first cigarette. You would be wise to learn from their mistake if you have not yet started smoking.
>
> According to the U.S. Department of Health and Human Services, 16 million people in the United States presently suffer from a smoking-related condition and nearly nine million suffer from a serious smoking-related illness. According to the Centers for Disease Control and Prevention (CDC), tobacco products cause nearly six million deaths per year. This number is projected to rise to over eight million deaths by 2030. Smokers, on average, die ten years earlier than their nonsmoking peers.
>
> In the United States, local, state, and federal governments typically tax tobacco products, which leads to high prices. Nicotine users who struggle with addiction sometimes pay more for a pack of cigarettes than for a few gallons of gas. Additionally, smokers tend to stink. The smell of smoke is all-consuming and creates a pervasive nastiness. Smokers also risk staining their teeth and fingers with yellow residue from the tar.
>
> Smoking is deadly, expensive, and socially unappealing. Clearly, smoking is not worth the risks.

12. Which of the following statements most accurately summarizes the passage?
    a. Tobacco is less healthy than many alternatives.
    b. Tobacco is deadly, expensive, and socially unappealing, and smokers would be much better off kicking the addiction.
    c. In the United States, local, state, and federal governments typically tax tobacco products, which leads to high prices.
    d. Tobacco products shorten smokers' lives by ten years and kill more than six million people per year.

13. The author would be most likely to agree with which of the following statements?
    a. Smokers should only quit cold turkey and avoid all nicotine cessation devices.
    b. Other substances are more addictive than tobacco.
    c. Smokers should quit for whatever reason that gets them to stop smoking.
    d. People who want to continue smoking should advocate for a reduction in tobacco product taxes.

14. Which of the following represents an opinion statement on the part of the author?
    a. According to the Centers for Disease Control and Prevention (CDC), tobacco products cause nearly six million deaths per year.
    b. Nicotine users who struggle with addiction sometimes pay more for a pack of cigarettes than a few gallons of gas.
    c. They also risk staining their teeth and fingers with yellow residue from the tar.
    d. Additionally, smokers tend to stink. The smell of smoke is all-consuming and creates a pervasive nastiness.

The next three questions are based on the following passage.

Christopher Columbus is often credited for discovering America. This is incorrect. First, it is impossible to "discover" something where people already live; however, Christopher Columbus did explore places in the New World that were previously untouched by Europe, so the term “explorer” would be more accurate. Another correction must be made, as well: Christopher Columbus was not the first European explorer to reach the present-day Americas! Rather, it was Leif Erikson who first came to the New World and contacted the natives, nearly five hundred years before Christopher Columbus.

Leif Erikson, the son of Erik the Red (a famous Viking outlaw and explorer in his own right), was born in either 970 or 980, depending on which historian you seek. His own family, though, did not raise Leif, which was a Viking tradition. Instead, one of Erik's prisoners taught Leif reading and writing, languages, sailing, and weaponry. At age 12, Leif was considered a man and returned to his family. He killed a man during a dispute shortly after his return, and the council banished the Erikson clan to Greenland.

In 999, Leif left Greenland and traveled to Norway where he would serve as a guard to King Olaf Tryggvason. It was there that he became a convert to Christianity. Leif later tried to return home with the intention of taking supplies and spreading Christianity to Greenland, however his ship was blown off course and he arrived in a strange new land: present day Newfoundland, Canada.

When he finally returned to his adopted homeland Greenland, Leif consulted with a merchant who had also seen the shores of this previously unknown land we now know as Canada. The son of the legendary Viking explorer then gathered a crew of 35 men and set sail. Leif became the first European to touch foot in the New World as he explored present-day Baffin Island and Labrador, Canada. His crew called the land Vinland since it was plentiful with grapes.

During their time in present-day Newfoundland, Leif's expedition made contact with the natives whom they referred to as Skraelings (which translates to "wretched ones" in Norse). There are several secondhand accounts of their meetings. Some contemporaries described trade between the peoples. Other accounts describe clashes where the Skraelings defeated the Viking explorers with long spears, while still others claim the Vikings dominated the natives. Regardless of the circumstances, it seems that the Vikings made contact of some kind. This happened around 1000, nearly five hundred years before Columbus famously sailed the ocean blue.

Eventually, in 1003, Leif set sail for home and arrived at Greenland with a ship full of timber. In 1020, seventeen years later, the legendary Viking died. Many believe that Leif Erikson should receive more credit for his contributions in exploring the New World.

15. Which of the following is an opinion, rather than historical fact, expressed by the author?
    a. Leif Erikson was definitely the son of Erik the Red; however, historians debate the year of his birth.
    b. Leif Erikson's crew called the land Vinland since it was plentiful with grapes.
    c. Leif Erikson deserves more credit for his contributions in exploring the New World.
    d. Leif Erikson explored the Americas nearly five hundred years before Christopher Columbus.

16. Which of the following most accurately describes the author's main conclusion?
    a. Leif Erikson is a legendary Viking explorer.
    b. Leif Erikson deserves more credit for exploring America hundreds of years before Columbus.
    c. Spreading Christianity motivated Leif Erikson's expeditions more than any other factor.
    d. Leif Erikson contacted the natives nearly five hundred years before Columbus.

17. Which of the following can be logically inferred from the passage?
    a. The Vikings disliked exploring the New World.
    b. Leif Erikson's banishment from Iceland led to his exploration of present-day Canada.
    c. Leif Erikson never shared his stories of exploration with the King of Norway.
    d. Historians have difficulty definitively pinpointing events in the Vikings' history

This article discusses the famous poet and playwright William Shakespeare. Read it and answer questions 18-21.

People who argue that William Shakespeare is not responsible for the plays attributed to his name are known as anti-Stratfordians (from the name of Shakespeare's birthplace, Stratford-upon-Avon). The most common anti-Stratfordian claim is that William Shakespeare simply was not educated enough or from a high enough social class to have written plays overflowing with references to such a wide range of subjects like history, the classics, religion, and international culture. William Shakespeare was the son of a glove-maker, he only had a basic grade school education, and he never set foot outside of England—so how could he have produced plays of such sophistication and imagination? How could he have written in such detail about historical figures and events, or about different cultures and locations around Europe? According to anti-Stratfordians, the depth of knowledge contained in Shakespeare's plays suggests a well-traveled writer from a wealthy background with a university education, not a countryside writer like Shakespeare. But in fact, there is not much substance to such speculation, and most anti-Stratfordian arguments can be refuted with a little background about Shakespeare's time and upbringing.

First of all, those who doubt Shakespeare's authorship often point to his common birth and brief education as stumbling blocks to his writerly genius. Although it is true that Shakespeare did not come from a noble class, his father was a very *successful* glove-maker and his mother was from a very wealthy land-owning family—so while Shakespeare may have had a country upbringing, he was certainly from a well-off family and would have been educated accordingly. Also, even though he did not attend university, grade school education in Shakespeare's time was actually quite rigorous and exposed students to classic drama through writers like Seneca and Ovid. It is not unreasonable to believe that Shakespeare received a very solid foundation in poetry and literature from his early schooling.

Next, anti-Stratfordians tend to question how Shakespeare could write so extensively about countries and cultures he had never visited before (for instance, several of his most famous works like *Romeo and Juliet* and *The Merchant of Venice* were set in Italy, on the opposite side of Europe!). But again, this criticism does not hold up under scrutiny. For one thing, Shakespeare was living in London, a bustling metropolis of international trade, the most populous city in England, and a political and cultural hub of Europe. In the daily crowds of people, Shakespeare would certainly have been able to meet travelers from other countries and hear firsthand accounts of life in their home country. And, in addition to the influx of information from world travelers, this was also the age of the printing press, a jump in technology that made it possible to print and circulate books much more easily than in the past. This also allowed for a freer flow of information across different countries, allowing people to read about life and ideas from throughout Europe. One needn't travel the continent in order to learn and write about its culture.

18. Which sentence contains the author's thesis?
    a. People who argue that William Shakespeare is not responsible for the plays attributed to his name are known as anti-Stratfordians.
    b. But in fact, there is not much substance to such speculation, and most anti-Stratfordian arguments can be refuted with a little background about Shakespeare's time and upbringing.
    c. It is not unreasonable to believe that Shakespeare received a very solid foundation in poetry and literature from his early schooling.
    d. Next, anti-Stratfordians tend to question how Shakespeare could write so extensively about countries and cultures he had never visited before.

19. In the first paragraph, "How could he have written in such detail about historical figures and events, or about different cultures and locations around Europe?" is an example of which of the following?
    a. Hyperbole
    b. Onomatopoeia
    c. Rhetorical question
    d. Appeal to authority

20. How does the author respond to the claim that Shakespeare was not well-educated because he did not attend university?
    a. By insisting upon Shakespeare's natural genius.
    b. By explaining grade school curriculum in Shakespeare's time.
    c. By comparing Shakespeare with other uneducated writers of his time.
    d. By pointing out that Shakespeare's wealthy parents probably paid for private tutors.

21. The word "bustling" in the third paragraph most nearly means which of the following?
    a. Busy
    b. Foreign
    c. Expensive
    d. Undeveloped

The next article is for questions 22-24.

The Myth of Head Heat Loss

It has recently been brought to my attention that most people believe that 75% of your body heat is lost through your head. I had certainly heard this before, and am not going to attempt to say I didn't believe it when I first heard it. It is natural to be gullible to anything said with enough authority. But the "fact" that the majority of your body heat is lost through your head is a lie.

Let me explain. Heat loss is proportional to surface area exposed. An elephant loses a great deal more heat than an anteater because it has a much greater surface area than an anteater. Each cell has mitochondria that produce energy in the form of heat, and it takes a lot more energy to run an elephant than an anteater.

So, each part of your body loses its proportional amount of heat in accordance with its surface area. The human torso probably loses the most heat, though the legs lose a significant amount as well. Some people have asked, "Why does it feel so much warmer when you cover your head than when you don't?" Well, that's because your head, because it is not clothed, is losing a lot of heat while the clothing on the rest of your body provides insulation. If you went outside with a hat and pants but no shirt, not only would you look silly, but your heat loss would be significantly greater because so much more of you would be exposed. So, if given the choice to cover your chest or your head in the cold, choose the chest. It could save your life.

22. Why does the author compare elephants and anteaters?
    a. To express an opinion.
    b. To give an example that helps clarify the main point.
    c. To show the differences between them.
    d. To persuade why one is better than the other.

23. Which of the following best describes the tone of the passage?
    a. Harsh
    b. Angry
    c. Casual
    d. Indifferent

24. The author appeals to which branch of rhetoric to prove their case?
    a. Factual evidence
    b. Emotion
    c. Ethics and morals
    d. Author qualification

Questions 25-30 are based upon the following passage:

This excerpt is an adaptation of Robert Louis Stevenson's *The Strange Case of Dr. Jekyll and Mr. Hyde.*

"Did you ever come across a protégé of his—one Hyde?" He asked.

"Hyde?" repeated Lanyon. "No. Never heard of him. Since my time."

That was the amount of information that the lawyer carried back with him to the great, dark bed on which he tossed to and fro until the small hours of the morning began to grow large. It was a night of little ease to his toiling mind, toiling in mere darkness and besieged by questions.

Six o'clock struck on the bells of the church that was so conveniently near to Mr. Utterson's dwelling, and still he was digging at the problem. Hitherto it had touched him on the intellectual side alone; but now his imagination also was engaged, or rather enslaved; and as he lay and tossed in the gross darkness of the night in the curtained room, Mr. Enfield's tale went by before his mind in a scroll of lighted pictures. He would be aware of the great field of lamps in a nocturnal city; then of the figure of a man walking swiftly; then of a child running from the doctor's; and then these met, and that human Juggernaut trod the child down and passed on regardless of her screams. Or else he would see a room in a rich house, where his friend lay asleep, dreaming and smiling at his dreams; and then the door of that room would be opened, the curtains of the bed plucked apart, the sleeper recalled, and, lo! There would stand by his side a figure to whom power was given, and even at that dead hour he must rise and do its bidding. The figure in these two phrases haunted the lawyer all night; and if at anytime he dozed over, it was but to see it glide more stealthily through sleeping houses, or move the more swiftly, and still the more smoothly, even to dizziness, through wider labyrinths of lamplighted city, and at every street corner crush a child and leave her screaming. And still the figure had no face by which he might know it; even in his dreams it had no face, or one that baffled him and melted before his eyes; and thus there it was that there sprung up and grew apace in the lawyer's mind a singularly strong, almost an inordinate, curiosity to behold the features of the real Mr. Hyde. If he could but once set eyes on him, he thought the mystery would lighten and perhaps roll altogether away, as was the habit of mysterious things when well examined. He might see a reason for his friend's strange preference or bondage, and even for the startling clauses of the will. And at least it would be a face worth seeing: the face of a man who was without bowels of mercy: a face which had but to show itself to raise up, in the mind of the unimpressionable Enfield, a spirit of enduring hatred.

From that time forward, Mr. Utterson began to haunt the door in the by street of shops. In the morning before office hours, at noon when business was plenty of time scarce, at night under the face of the full city moon, by all lights and at all hours of solitude or concourse, the lawyer was to be found on his chosen post.

"If he be Mr. Hyde," he had thought, "I should be Mr. Seek."

25. What is the purpose of the use of repetition in the following passage?

It was a night of little ease to his toiling mind, toiling in mere darkness and besieged by questions.

a. It serves as a demonstration of the mental state of Mr. Lanyon.
b. It is reminiscent of the church bells that are mentioned in the story.
c. It mimics Mr. Utterson's ambivalence.
d. It emphasizes Mr. Utterson's anguish in failing to identify Hyde's whereabouts.

26. What is the setting of the story in this passage?
    a. In the city
    b. On the countryside
    c. In a jail
    d. In a mental health facility

27. What can one infer about the meaning of the word "Juggernaut" from the author's use of it in the passage?
    a. It is an apparition that appears at daybreak.
    b. It scares children.
    c. It is associated with space travel.
    d. Mr. Utterson finds it soothing.

28. What is the definition of the word *haunt* in the following passage?

    From that time forward, Mr. Utterson began to haunt the door in the by street of shops. In the morning before office hours, at noon when business was plenty of time scarce, at night under the face of the full city moon, by all lights and at all hours of solitude or concourse, the lawyer was to be found on his chosen post.

    a. To levitate
    b. To constantly visit
    c. To terrorize
    d. To daunt

29. The phrase *labyrinths of lamplighted city* contains an example of what?
    a. Hyperbole
    b. Simile
    c. Juxtaposition
    d. Alliteration

30. What can one reasonably conclude from the final comment of this passage?

    "If he be Mr. Hyde," he had thought, "I should be Mr. Seek."

    a. The speaker is considering a name change.
    b. The speaker is experiencing an identity crisis.
    c. The speaker has mistakenly been looking for the wrong person.
    d. The speaker intends to continue to look for Hyde.

Questions 31-34 are based upon the following passage:

> This excerpt is adapted from "What to the Slave is the Fourth of July?" Rochester, New York July 5, 1852.
>
> Fellow citizens—Pardon me, and allow me to ask, why am I called upon to speak here today? What have I, or those I represent, to do with your national independence? Are the great principles of political freedom and of natural justice, embodied in that Declaration of Independence, extended to us? And am I therefore called upon to bring our humble offering to the national altar, and to confess the benefits, and express devout gratitude for the blessings, resulting from your independence to us?

Would to God, both for your sakes and ours, ours that an affirmative answer could be truthfully returned to these questions! Then would my task be light, and my burden easy and delightful. For who is there so cold that a nation's sympathy could not warm him? Who so obdurate and dead to the claims of gratitude that would not thankfully acknowledge such priceless benefits? Who so stolid and selfish, that would not give his voice to swell the hallelujahs of a nation's jubilee, when the chains of servitude had been torn from his limbs? I am not that man. In a case like that, the dumb my eloquently speak, and the lame man leap as an hart.

But, such is not the state of the case. I say it with a sad sense of the disparity between us. I am not included within the pale of this glorious and anniversary. Oh pity! Your high independence only reveals the immeasurable distance between us. The blessings in which you this day rejoice, I do not enjoy in common. The rich inheritance of justice, liberty, prosperity, and independence, bequeathed by your fathers, is shared by *you*, not by *me*. This Fourth of July is *yours,* not *mine*. You may rejoice, *I* must mourn. To drag a man in fetters into the grand illuminated temple of liberty, and call upon him to join you in joyous anthems, were inhuman mockery and sacrilegious irony. Do you mean, citizens, to mock me, by asking me to speak today? If so there is a parallel to your conduct. And let me warn you that it is dangerous to copy the example of a nation whose crimes, towering up to heaven, were thrown down by the breath of the Almighty, burying that nation and irrecoverable ruin! I can today take up the plaintive lament of a peeled and woe-smitten people.

By the rivers of Babylon, there we sat down. Yea! We wept when we remembered Zion. We hanged our harps upon the willows in the midst thereof. For there, they that carried us away captive, required of us a song; and they who wasted us required of us mirth, saying, "Sing us one of the songs of Zion." How can we sing the Lord's song in a strange land? If I forget thee, O Jerusalem, let my right hand forget her cunning. If I do not remember thee, let my tongue cleave to the roof of my mouth.

31. What is the tone of the first paragraph of this passage?
    a. Exasperated
    b. Inclusive
    c. Contemplative
    d. Nonchalant

32. Which word CANNOT be used synonymously with the term *obdurate* as it is conveyed in the text below?

    Who so obdurate and dead to the claims of gratitude, that would not thankfully acknowledge such priceless benefits?

    a. Steadfast
    b. Stubborn
    c. Contented
    d. Unwavering

33. What is the central purpose of this text?
    a. To demonstrate the author's extensive knowledge of the Bible
    b. To address the hypocrisy of the Fourth of July holiday
    c. To convince wealthy landowners to adopt new holiday rituals
    d. To explain why minorities often relished the notion of segregation in government institutions

34. Which statement serves as evidence for the question above?
    a. By the rivers of Babylon, there we sat down.
    b. Fellow citizens—Pardon me, and allow me to ask, why am I called upon to speak here today?
    c. I can today take up the plaintive lament of a peeled and woe-smitten people.
    d. The rich inheritance of justice, liberty, prosperity, and independence, bequeathed by your fathers, is shared by *you*, not by *me*.

Directions for questions 35–44

For the questions that follow, two underlined sentences are followed by a question or statement. Read the sentences, then choose the best answer to the question or the best completion of the statement.

35. The NBA draft process is based on a lottery system among the teams who did not make the playoffs in the previous season to determine draft order. Only the top three draft picks are determined by the lottery.

What does the *second sentence* do?

a. It contradicts the first.
b. It supports the first.
c. It restates information from the first.
d. It offers a solution.

36. While many people use multiple social media sites, Facebook remains the most popular with more than one billion users. Instagram is rising in popularity and, with 100 million users, is now the most-used social media site.

What does the *second sentence* do?

a. It expands on the first.
b. It contradicts the first.
c. It supports the first.
d. It proposes a solution.

37. There are eight different phases of the moon, from new moon to new moon. One of the eight different moon phases is first quarter, commonly called a half moon.

What does the *second sentence* do?

a. It provides an example.
b. It contradicts the first.
c. It states an effect.
d. It offers a solution.

38. The terror attacks of September 11, 2001 have had many lasting effects on the United States. The Department of Homeland Security was created in late September 2001 in response to the terror attacks and became an official cabinet-level department in November of 2002.

What does the *second sentence* do?

a. It contradicts the first.
b. It restates the information from the first.
c. It states an effect.
d. It makes a contrast.

39. Annuals are plants that complete the life cycle in a single growing season. Perennials are plants that complete the life cycle in many growing seasons, dying in the winter and coming back each spring.

What does the *second sentence* do?

a. It makes a contrast.
b. It disputes the first sentence.
c. It provides an example.
d. It states an effect.

40. Personal computers can be subject to viruses and malware, which can lead to slow performance, loss of files, and overheating. Antivirus software is often sold along with a new PC to protect against viruses and malware.

What does the *second sentence* do?

a. It makes a contrast.
b. It provides an example.
c. It restates the information from the first.
d. It offers a solution.

41. Many companies tout their chicken as cage-free because the chickens are not confined to small wire cages. However, cage-free chickens are often crammed into buildings with thousands of other birds and never go outside in their short lifetime.

What does the *second sentence* do?

a. It offers a solution.
b. It provides an example.
c. It disputes the first sentence.
d. It states an effect.

42. Common core standards do not include the instruction of cursive handwriting. The next generation of students will not be able to read or write cursive handwriting.

What does the *second sentence* do?

a. It offers a solution.
b. It states an effect.
c. It contradicts the first sentence.
d. It restates the first sentence.

43. Air travel has changed significantly in the last ten years. Airlines are now offering pay-as-you-go perks, including no baggage fees, seat selection, and food and drinks on the flight to keep costs low.

What does the *second sentence* do?

a. It states effects.
b. It provides examples.
c. It disputes the first sentence.
d. It offers solutions.

44. <u>Many people are unaware that fragrances and other chemicals in their favorite products are causing skin reactions and allergies. Surprisingly, many popular products contain ingredients that can cause skin allergies.</u>

What does the *second sentence* do?

a. It restates the first sentence.
b. It provides examples.
c. It contradicts the first sentence.
d. It provides solutions.

Answer Explanations

1. D: The passage directly states that the larger sensor is the main difference between the two cameras. Choices *A* and *B* may be true, but these answers do not identify the major difference between the two cameras. Choice *C* states the opposite of what the paragraph suggests is the best option for amateur photographers, so it is incorrect.

2. D: An actuary assesses risks and sets insurance premiums. While an actuary does work in insurance, the passage does not suggest that actuaries have any affiliation with hospitalists or working in a hospital, so all other choices are incorrect.

3. A: The passage focuses mainly on the problems of hard water. Choice *B* is incorrect because calcium is not good for pipes and hard surfaces. The passage does not say anything about whether water softeners are easy to install, so Choice *C* is incorrect. Choice *D* is also incorrect because the passage does offer other solutions besides vinegar.

4. C: The main point of this paragraph is that parents need to change their poor behavior at their kids' sporting events. Choice *A* is incorrect because the coaches' behavior is not mentioned in the paragraph. Choice *B* suggests that sports are bad for kids, when the paragraph is about parents' behavior, so it is incorrect. While Choice *D* may be true, it offers a specific solution to the problem, which the paragraph does not discuss.

5. B: The main point of this passage is to show how tornadoes form. Choice *A* is off base because while the passage does mention that tornadoes are dangerous, it is not the main focus of the passage. While thunderstorms are mentioned, they are not compared to tornadoes, so Choice *C* is incorrect. Choice *D* is incorrect because the passage does not discuss what to do in the event of a tornado.

6. C: The purpose of this passage is to explain how the digestive system works. Choice *A* focuses only on the liver, which is a small part of the process and not the focus of the paragraph. Choice *B* is off-track because the passage does not mention healthy foods. Choice *D* only focuses on one part of the digestive system.

7. D: The main point of this passage is to define osteoporosis. Choice *A* is incorrect because the passage does not list ways that people contract osteoporosis. Choice *B* is incorrect because the passage does not mention any treatment options. While the passage does briefly mention prevention, it does not explain how, so Choice *C* is incorrect.

8. D: The passage compares Disney cruises with Disney parks. It does not discuss how to book a cruise, so Choice *A* is incorrect. Choice *B* is incorrect because though the passage does mention some of the park attractions, it is not the main point. The passage does not mention the cost of either option, so Choice *C* is incorrect.

9. B: The point of this passage is to show what drowning looks like. Choice *A* is incorrect because while drowning is a danger of swimming, the passage doesn't include any other dangers. The passage is not intended for lifeguards specifically, but for a general audience, so Choice *C* is incorrect. There are a few signs of drowning, but the passage does not compare them; thus, *D* is incorrect.

10. B: The scientist and politician largely disagree, but the question asks for a point where the two are in agreement. The politician would not concur that burning fossil fuels causes global temperatures to rise;

thus, Choice *A* is wrong. The politician also would not agree with Choice *C* suggesting that countries must revisit their energy policies. By inference from the given information, the scientist would likely not concur that earth's climate naturally goes through warming and cooling cycles; so Choice *D* is incorrect. However, both the scientist and politician would agree that global temperatures are increasing. The reason for this is in dispute. The politician thinks it is part of the earth's natural cycle; the scientist thinks it is from the burning of fossil fuels. However, both acknowledge an increase, so Choice *B* is the correct answer.

11. C: We are looking for an inference—a conclusion that is reached on the basis of evidence and reasoning—from the passage that will likely explain why the famous children's author did not achieve her usual success with the new genre (despite the book's acclaim). Choice *A* is wrong because the statement is false according to the passage. Choice *B* is wrong because, although the passage says the author has a graduate degree on the subject, it would be an unrealistic leap to infer that she is the foremost expert on Antebellum America. Choice *D* is wrong because there is nothing in the passage to lead us to infer that people generally prefer a children's series to historical fiction. In contrast, Choice *C* can be logically inferred since the passage speaks of the great success of the children's series and the declaration that the fame of the author's name causes the children's books to "fly off the shelves." Thus, she did not receive any bump from her name since she published the historical novel under a pseudonym, and Choice *C* is correct.

12. B: The author is opposed to tobacco. He cites disease and deaths associated with smoking. He points to the monetary expense and aesthetic costs. Choice *A* is incorrect because alternatives to smoking are not even addressed in the passage. Choice *C* is incorrect because it does not summarize the passage but rather is just a premise. Choice *D* is incorrect because, while these statistics are a premise in the argument, they do not represent a summary of the piece. Choice *B* is the correct answer because it states the three critiques offered against tobacco and expresses the author's conclusion.

13. C: We are looking for something the author would agree with, so it should be anti-smoking or an argument in favor of quitting smoking. Choice *A* is incorrect because the author does not speak against means of cessation. Choice *B* is incorrect because the author does not reference other substances but does speak of how addictive nicotine, a drug in tobacco, is. Choice *D* is incorrect because the author would not encourage reducing taxes to encourage a reduction of smoking costs, thereby helping smokers to continue the habit. Choice *C* is correct because the author is attempting to persuade smokers to quit smoking.

14. D: Here, we are looking for an opinion of the author's rather than a fact or statistic. Choice *A* is incorrect because quoting statistics from the Centers of Disease Control and Prevention is stating facts, not opinions. Choice *B* is incorrect because it expresses the fact that cigarettes sometimes cost more than a few gallons of gas. It would be an opinion if the author said that cigarettes were not affordable. Choice *C* is incorrect because yellow stains are a known possible adverse effect of smoking. Choice *D* is correct as an opinion because smell is subjective. Some people might like the smell of smoke, they might not have working olfactory senses, and/or some people might not find the smell of smoke akin to "pervasive nastiness," so this is the expression of an opinion. Thus, Choice *D* is the correct answer.

15. C: Choice *A* is incorrect because it describes facts: Leif Erikson was the son of Erik the Red and historians debate Leif's date of birth. These are not opinions. Choice *B* is incorrect; that Erikson called the land Vinland is a verifiable fact as is Choice *D* because he did contact the natives almost 500 years before Columbus. Choice *C* is the correct answer because it is the author's opinion that Erikson deserves more credit. That, in fact, is his conclusion in the piece, but another person could argue that Columbus or another explorer deserves more credit for opening up the New World to exploration. Rather than being an incontrovertible fact, it is a subjective value claim.

16. B: Choice *A* is incorrect because the author aims to go beyond describing Erikson as a mere legendary Viking. Choice *C* is incorrect because the author does not focus on Erikson's motivations, let alone name the spreading of Christianity as his primary objective. Choice *D* is incorrect because it is a premise that Erikson contacted the natives 500 years before Columbus, which is simply a part of supporting the author's conclusion. Choice *B* is correct because, as stated in the previous answer, it accurately identifies the author's statement that Erikson deserves more credit than he has received for being the first European to explore the New World.

17. D: Choice *A* is incorrect because the author never addresses the Vikings' state of mind or emotions. Choice *B* is incorrect because the author does not elaborate on Erikson's exile and whether he would have become an explorer if not for his banishment. Choice *C* is incorrect because there is not enough information to support this premise. It is unclear whether Erikson informed the King of Norway of his finding. Although it is true that the King did not send a follow-up expedition, he could have simply chosen not to expend the resources after receiving Erikson's news. It is not possible to logically infer whether Erikson told him. Choice *D* is correct because there are two examples—Leif Erikson's date of birth and what happened during the encounter with the natives—of historians having trouble pinning down important details in Viking history.

18. B: But in fact, there is not much substance to such speculation, and most anti-Stratfordian arguments can be refuted with a little background about Shakespeare's time and upbringing. The thesis is a statement that contains the author's topic and main idea. The main purpose of this article is to use historical evidence to provide counterarguments to anti-Stratfordians. Choice *A* is simply a definition; Choice *C* is a supporting detail, not a main idea; and Choice *D* represents an idea of anti-Stratfordians, not the author's opinion.

19. C: It is an example of a rhetorical question. This requires readers to be familiar with different types of rhetorical devices. A rhetorical question is a question that is asked not to obtain an answer but to encourage readers to more deeply consider an issue.

20. B: By explaining grade school curriculum in Shakespeare's time. This question asks readers to refer to the organizational structure of the article and demonstrate understanding of how the author provides details to support their argument. This particular detail can be found in the second paragraph: "even though he did not attend university, grade school education in Shakespeare's time was actually quite rigorous."

21. A: It most closely means busy. This is a vocabulary question that can be answered using context clues. Other sentences in the paragraph describe London as "the most populous city in England" filled with "crowds of people," giving an image of a busy city full of people. Choice *B* is incorrect because London was in Shakespeare's home country, not a foreign one. Choice *C* is not mentioned in the passage. Choice *D* is not a good answer choice because the passage describes how London was a popular and important city, probably not an undeveloped one.

22. B: Choice *B* is correct because the author is trying to demonstrate the main idea, which is that heat loss is proportional to surface area, and so they compare two animals with different surface areas to clarify the main point. Choice *A* is incorrect because the author uses elephants and anteaters to prove a point, that heat loss is proportional to surface area, not to express an opinion. Choice *C* is incorrect because though the author does use them to show differences, they do so in order to give examples that prove the above points, so Choice *C* is not the best answer. Choice *D* is incorrect because there is no language to indicate favoritism between the two animals.

23. C: Because of the way that the author addresses the reader, and also the colloquial language that the author uses (i.e., "let me explain," "so," "well," didn't," "you would look silly," etc.), *C* is the best answer because it has a much more casual tone than the usual informative article. Choice A may be a tempting choice because the author says the "fact" that most of one's heat is lost through their head is a "lie," and that someone who does not wear a shirt in the cold looks silly, but it only happens twice within all the diction of the passage and it does not give an overall tone of harshness. *B* is incorrect because again, while not necessarily nice, the language does not carry an angry charge. The author is clearly not indifferent to the subject because of the passionate language that they use, so *D* is incorrect.

24. A: The author gives logical examples and reasons in order to prove that most of one's heat is not lost through their head; therefore, *A* is correct. *B* is incorrect because there is not much emotionally charged language in this selection, and even the small amount present is greatly outnumbered by the facts and evidence. *C* is incorrect because there is no mention of ethics or morals in this selection. *D* is incorrect because the author never qualifies himself as someone who has the authority to be writing on this topic.

25. D: It emphasizes Mr. Utterson's anguish in failing to identify Hyde's whereabouts. Context clues indicate that Choice *D* is correct because the passage provides great detail of Mr. Utterson's feelings about locating Hyde. Choice *A* does not fit because there is no mention of Mr. Lanyon's mental state. Choice *B* is incorrect; although the text does make mention of bells, Choice *B* is not the *best* answer overall. Choice *C* is incorrect because the passage clearly states that Mr. Utterson was determined, not unsure.

26. A: In the city. The word *city* appears in the passage several times, thus establishing the location for the reader.

27. B: It scares children. The passage states that the Juggernaut causes the children to scream. Choices *A* and *D* don't apply because the text doesn't mention either of these instances specifically. Choice *C* is incorrect because there is nothing in the text that mentions space travel.

28. B: To constantly visit. The mention of *morning*, *noon*, and *night* make it clear that the word *haunt* refers to frequent appearances at various locations. Choice *A* doesn't work because the text makes no mention of levitating. Choices *C* and *D* are not correct because the text makes mention of Mr. Utterson's anguish and disheartenment because of his failure to find Hyde but does not make mention of Mr. Utterson's feelings negatively affecting anyone else.

29. D: This is an example of alliteration. Choice *D* is the correct answer because of the repetition of the *L*-words. Hyperbole is an exaggeration, so Choice *A* doesn't work. No comparison is being made, so no simile or juxtaposition is being used, thus eliminating Choices B and C.

30. D: The speaker intends to continue to look for Hyde. Choices *A* and *B* are not possible answers because the text doesn't refer to any name changes or an identity crisis, despite Mr. Utterson's extreme obsession with finding Hyde. The text also makes no mention of a mistaken identity when referring to Hyde, so Choice *C* is also incorrect.

31. A: The tone is exasperated. While contemplative is an option because of the inquisitive nature of the text, Choice *A* is correct because the speaker is frustrated by the thought of being included when he felt that the fellow members of his race were being excluded. The speaker is not nonchalant, nor accepting of the circumstances which he describes.

32. C: Choice *C*, *contented*, is the only word that has different meaning. Furthermore, the speaker expresses objection and disdain throughout the entire text.

33. B: To address the hypocrisy of the Fourth of July holiday. While the speaker makes biblical references, it is not the main focus of the passage, thus eliminating Choice *A* as an answer. The passage also makes no mention of wealthy landowners and doesn't speak of any positive response to the historical events, so Choices *C* and *D* are not correct.

34: D: Choice *D* is the correct answer because it clearly makes reference to justice being denied.

35. B: The information in the second sentence further explains the draft process and thus supports the first sentence. It does not contradict the first sentence, so *A* is incorrect. Choice *C* and *D* are incorrect because the second sentence does not restate or offer a solution to the first.

36. B: The first sentence identifies Facebook as the most popular social media site with 1 billion users. The second sentence states that Instagram only has 100 million users, but is the most used. This contradicts the original sentence, so all other answers are incorrect.

37. A: The first sentence states that there are eight phases to the moon cycle. The second sentence discusses first quarter, which is one of the phases of the moon. Therefore, the second sentence provides an example of the first sentence.

38. C: The first sentence discusses the effects of the terror attacks of September 11, 2001. The second sentence states that the Department of Homeland Security was created in response to the terror attacks, so it states an effect of the first sentence.

39. A: The first sentence describes the life cycle of annuals. The second sentence describes the life cycle of perennials, making a contrast between the way annuals grow and the way perennials grow.

40. D: The first sentence describes how viruses can affect a PC. The second sentence offers a solution to the problem of viruses and malware on a PC.

41. C: The first sentence describes cage-free chickens as not being confined to a cage, suggesting they are treated humanely. The second sentence disputes the first sentence, showing that cage-free chickens are inhumanely confined to a larger area with many other chickens, never seeing the outdoors.

42. B: The first sentence states that schools are no longer teaching cursive handwriting. The second sentence shows that as an effect of the first sentence, students will no longer be able to read or write cursive handwriting.

43. B: The first sentence states that air travel has changed in the last decade. The second sentence provides examples of the changes that have occurred.

44. A: The first sentence discusses how fragrances and other chemicals in products can cause skin reactions. The second sentence states that many products contain ingredients that cause skin allergies, restating the same information from the first sentence.

Sentence Skills Test

Sentence Correction

In this section of the test, each example contains a sentence with an underlined portion. The multiple-choice answers will offer reworded versions of the underlined part of the sentence. The first answer choice will repeat the original sentence, and the others will offer different options for how to re-word the sentence.

The goal of these questions is to test an individual's ability to recognize the most effective way to express something, using correct grammar, word choice, and sentence structure. The correct answer will be grammatically-correct, with a clear, concise flow.

Construction Shift

These multiple-choice questions ask the test-taker to rewrite the original sentence so that it is more correct in terms of grammar, word choice, and sentence structure. These questions will provide the beginning of the new sentence and then offer four options for ending the sentence. The goal is to select the answer choice that makes the best structured sentence, while retaining the meaning of the original sentence. In most cases, when the sentence is rewritten, it will entail changing a dependent clause to an independent clause, or an independent clause to a dependent clause.

Answers should be chosen carefully, ensuring that they adhere to the rules of grammar and standard English. The best choice will be the one that is correct, concise, and clear, while also retaining the meaning of the original sentence.

A helpful strategy is to predict the sentence ending before looking at the answer choices. This approach helps in selecting the one that makes the most sense. Reading the answer choices first might be confusing and make the choice more difficult. In these questions, the simplest answer is often the best answer.

Types of Sentences

There isn't an overabundance of absolutes in grammar, but here is one: every sentence in the English language falls into one of four categories.

- Declarative: a simple statement that ends with a period

  The price of milk per gallon is the same as the price of gasoline.

- Imperative: a command, instruction, or request that ends with a period

  Buy milk when you stop to fill up your car with gas.

- Interrogative: a question that ends with a question mark

  Will you buy the milk?

- Exclamatory: a statement or command that expresses emotions like anger, urgency, or surprise and ends with an exclamation mark

Buy the milk now!

Declarative sentences are the most common type, probably because they are comprised of the most general content, without any of the bells and whistles that the other three types contain. They are, simply, declarations or statements of any degree of seriousness, importance, or information.

Imperative sentences often seem to be missing a subject. The subject is there, though; it is just not visible or audible because it is *implied*. Look at the imperative example sentence.

Buy the milk when you fill up your car with gas.

You is the implied subject, the one to whom the command is issued. This is sometimes called *the understood you* because it is understood that *you* is the subject of the sentence.

Interrogative sentences—those that ask questions—are defined as such from the idea of the word *interrogation*, the action of questions being asked of suspects by investigators. Although that is serious business, interrogative sentences apply to all kinds of questions.

To exclaim is at the root of *exclamatory* sentences. These are made with strong emotions behind them. The only technical difference between a declarative or imperative sentence and an exclamatory one is the exclamation mark at the end. The example declarative and imperative sentences can both become an exclamatory one simply by putting an exclamation mark at the end of the sentences.

The price of milk per gallon is the same as the price of gasoline!
Buy milk when you stop to fill up your car with gas!

After all, someone might be really excited by the price of gas or milk, or they could be mad at the person that will be buying the milk! However, as stated before, exclamation marks in abundance defeat their own purpose! After a while, they begin to cause fatigue! When used only for their intended purpose, they can have their expected and desired effect.

Parts of Speech

Nouns

A noun is a person, place, thing, or idea. All nouns fit into one of two types, common or proper.

A *common noun* is a word that identifies any of a class of people, places, or things. Examples include numbers, objects, animals, feelings, concepts, qualities, and actions. *A, an,* or *the* usually precedes the common noun. These parts of speech are called *articles*. Here are some examples of sentences using nouns preceded by articles.

A building is under construction.
The girl would like to move to *the* city.

A *proper noun* (also called a *proper name*) is used for the specific name of an individual person, place, or organization. The first letter in a proper noun is capitalized. "My name is *Mary*." "I work for *Walmart*."

Nouns sometimes serve as adjectives (which themselves describe nouns), such as "hockey player" and "state government."

An abstract noun is an idea, state, or quality. It is something that can't be touched, such as happiness, courage, evil, or humor.

A concrete noun is something that can be experienced through the senses (touch, taste, hear, smell, see). Examples of concrete nouns are birds, skateboard, pie, and car.

A collective noun refers to a collection of people, places, or things that act as one. Examples of collective nouns are as follows: team, class, jury, family, audience, and flock.

Pronouns

A word used in place of a noun is known as a *pronoun*. Pronouns are words like *I, mine, hers,* and *us*.

Pronouns can be split into different classifications (see below) which make them easier to learn; however, it's not important to memorize the classifications.

- Personal pronouns: refer to people
- First person: we, I, our, mine
- Second person: you, yours
- Third person: he, them
- Possessive pronouns: demonstrate ownership (mine, his, hers, its, ours, theirs, yours)
- Interrogative pronouns: ask questions (what, which, who, whom, whose)
- Relative pronouns: include the five interrogative pronouns and others that are relative (whoever, whomever, that, when, where)
- Demonstrative pronouns: replace something specific (this, that, those, these)
- Reciprocal pronouns: indicate something was done or given in return (each other, one another)
- Indefinite pronouns: have a nonspecific status (anybody, whoever, someone, everybody, somebody)

Indefinite pronouns such as *anybody, whoever, someone, everybody*, and *somebody* command a singular verb form, but others such as *all, none,* and *some* could require a singular or plural verb form.

Antecedents

An *antecedent* is the noun to which a pronoun refers; it needs to be written or spoken before the pronoun is used. For many pronouns, antecedents are imperative for clarity. In particular, many of the personal, possessive, and demonstrative pronouns need antecedents. Otherwise, it would be unclear who or what someone is referring to when they use a pronoun like *he* or *this.*

Pronoun reference means that the pronoun should refer clearly to one, clear, unmistakable noun (the antecedent).

Pronoun-antecedent agreement refers to the need for the antecedent and the corresponding pronoun to agree in gender, person, and number. Here are some examples:

> The *kidneys* (plural antecedent) are part of the urinary system. *They* (plural pronoun) serve several roles."

The kidneys are part of the *urinary system* (singular antecedent). *It* (singular pronoun) is also known as the renal system.

Pronoun Cases

The subjective pronouns —*I, you, he/she/it, we, they,* and *who*—are the subjects of the sentence.

Example: *They* have a new house.

The objective pronouns—*me, you* (*singular*), *him/her, us, them,* and *whom*—are used when something is being done for or given to someone; they are objects of the action.

Example: The teacher has an apple for *us*.

The possessive pronouns—*mine, my, your, yours, his, hers, its, their, theirs, our,* and *ours*—are used to denote that something (or someone) belongs to someone (or something).

Example: It's *their* chocolate cake.
Even Better Example: It's *my* chocolate cake!

One of the greatest challenges and worst abuses of pronouns concerns *who* and *whom*. Just knowing the following rule can eliminate confusion. *Who* is a subjective-case pronoun used only as a subject or subject complement. *Whom* is only objective-case and, therefore, the object of the verb or preposition.

Who is going to the concert?

You are going to the concert with *whom*?

Hint: When using *who* or *whom*, think of whether someone would say *he* or *him*. If the answer is *he*, use *who*. If the answer is *him*, use *whom*. This trick is easy to remember because *he* and *who* both end in vowels, and *him* and *whom* both end in the letter *M*.

Verbs

The *verb* is the part of speech that describes an action, state of being, or occurrence.

A *verb* forms the main part of a predicate of a sentence. This means that the verb explains what the noun (which will be discussed shortly) is doing. A simple example is *time flies*. The verb *flies* explains what the action of the noun, *time*, is doing. This example is a *main* verb.

Helping (*auxiliary*) verbs are words like *have, do, be, can, may, should, must,* and *will.* "I *should* go to the store." Helping verbs assist main verbs in expressing tense, ability, possibility, permission, or obligation.

Particles are minor function words like *not, in, out, up,* or *down* that become part of the verb itself. "I might *not*."

Participles are words formed from verbs that are often used to modify a noun, noun phrase, verb, or verb phrase.

The *running* teenager collided with the cyclist.

Participles can also create compound verb forms.

He is *speaking*.

Verbs have five basic forms: the *base* form, the -*s* form, the -*ing* form, the *past* form, and the *past participle* form.

The *past* forms are either *regular* (*love/loved; hate/hated*) or *irregular* because they don't end by adding the common past tense suffix "-ed" (*go/went; fall/fell; set/set*).

Verb Forms

Shifting verb forms entails *conjugation*, which is used to indicate *tense, voice,* or *mood*.

Verb tense is used to show when the action in the sentence took place. There are several different verb tenses, and it is important to know how and when to use them. Some verb tenses can be achieved by changing the form of the verb, while others require the use of helping verbs (e.g., *is, was,* or *has*).

Present tense shows the action is happening currently or is ongoing:

> I walk to work every morning.
>
> She is stressed about the deadline.

Past tense shows that the action happened in the past or that the state of being is in the past:

> I walked to work yesterday morning.
>
> She was stressed about the deadline.

Future tense shows that the action will happen in the future or is a future state of being:

> I will walk to work tomorrow morning.
>
> She will be stressed about the deadline.

Present perfect tense shows action that began in the past, but continues into the present:

> I have walked to work all week.
>
> She has been stressed about the deadline.

Past perfect tense shows an action was finished before another took place:

> I had walked all week until I sprained my ankle.
>
> She had been stressed about the deadline until we talked about it.

Future perfect tense shows an action that will be completed at some point in the future:

> By the time the bus arrives, I will have walked to work already.

Voice

Verbs can be in the active or passive voice. When the subject completes the action, the verb is in *active voice*. When the subject receives the action of the sentence, the verb is in *passive voice*.

Active: Jamie ate the ice cream.

Passive: The ice cream was eaten by Jamie.

In active voice, the subject (*Jamie*) is the "do-er" of the action (*ate*). In passive voice, the subject *ice cream* receives the action of being eaten.

While passive voice can add variety to writing, active voice is the generally preferred sentence structure.

Mood

Mood is used to show the speaker's feelings about the subject matter. In English, there is *indicative mood, imperative mood,* and *subjective mood.*

Indicative mood is used to state facts, ask questions, or state opinions:

Bob will make the trip next week.

When can Bob make the trip?

Imperative mood is used to state a command or make a request:

Wait in the lobby.

Please call me next week.

Subjunctive mood is used to express a wish, an opinion, or a hope that is contrary to fact:

If I were in charge, none of this would have happened.

Allison wished she could take the exam over again when she saw her score.

Adjectives

Adjectives are words used to modify nouns and pronouns. They can be used alone or in a series and are used to further define or describe the nouns they modify.

Mark made us a delicious, four-course meal.

The words *delicious* and *four-course* are adjectives that describe the kind of meal Mark made.

Articles are also considered adjectives because they help to describe nouns. Articles can be general or specific. The three articles in English are: a, an, and the.

Indefinite articles (a, an) are used to refer to nonspecific nouns. The article *a* proceeds words beginning with consonant sounds, and the article *an* proceeds words beginning with vowel sounds.

A car drove by our house.

An alligator was loose at the zoo.

He has always wanted a ukulele. (The first *u* makes a *y* sound.)

Note that *a* and *an* should only proceed nonspecific nouns that are also singular. If a nonspecific noun is plural, it does not need a preceding article.

Alligators were loose at the zoo.

The *definite article (the)* is used to refer to specific nouns:

The car pulled into our driveway.

Note that *the* should proceed all specific nouns regardless of whether they are singular or plural.

The cars pulled into our driveway.

Comparative adjectives are used to compare nouns. When they are used in this way, they take on positive, comparative, or superlative form.

The *positive* form is the normal form of the adjective:

Alicia is tall.

The *comparative* form shows a comparison between two things:

Alicia is taller than Maria.

Superlative form shows comparison between more than two things:

Alicia is the tallest girl in her class.

Usually, the comparative and superlative can be made by adding *–er* and *–est* to the positive form, but some verbs call for the helping verbs *more* or *most*. Other exceptions to the rule include adjectives like *bad*, which uses the comparative *worse* and the superlative *worst*.

An adjective phrase is not a bunch of adjectives strung together, but a group of words that describes a noun or pronoun and, thus, functions as an adjective. Very happy is an adjective phrase; so are way too hungry and passionate about traveling.

Adverbs

Adverbs have more functions than adjectives because they modify or qualify verbs, adjectives, or other adverbs as well as word groups that express a relation of place, time, circumstance, or cause. Therefore, adverbs answer any of the following questions: *How, when, where, why, in what way, how often, how much, in what condition,* and/or *to what degree. How good looking is he? He is <u>very</u> handsome.*

Here are some examples of adverbs for different situations:

- how: quickly
- when: daily
- where: there
- in what way: easily
- how often: often
- how much: much
- in what condition: badly
- what degree: hardly

As one can see, for some reason, many adverbs end in *-ly.*

Adverbs do things like emphasize (*really, simply,* and *so*), amplify (*heartily, completely,* and *positively*), and tone down (*almost, somewhat,* and *mildly*).

Adverbs also come in phrases.

> *The dog ran as <u>though his life depended on it.</u>*

Prepositions

Prepositions are connecting words and, while there are only about 150 of them, they are used more often than any other individual groups of words. They describe relationships between other words. They are placed before a noun or pronoun, forming a phrase that modifies another word in the sentence. *Prepositional phrases* begin with a preposition and end with a noun or pronoun, the *object of the preposition. A pristine lake is <u>near the store</u> and <u>behind the bank</u>.*

Some commonly used prepositions are *about, after, anti, around, as, at, behind, beside, by, for, from, in, into, of, off, on, to,* and *with.*

Complex prepositions, which also come before a noun or pronoun, consist of two or three words such as *according to, in regards to,* and *because of.*

Conjunctions

Conjunctions are vital words that connect words, phrases, thoughts, and ideas. Conjunctions show relationships between components. There are two types:

Coordinating conjunctions are the primary class of conjunctions placed between words, phrases, clauses, and sentences that are of equal grammatical rank; the coordinating conjunctions are *for, and, nor, but, or, yet,* and *so*. A useful memorization trick is to remember that all the first letters of these conjunctions collectively spell the word fanboys.

> I need to go shopping, *but* I must be careful to leave enough money in the bank.
> She wore a black, red, *and* white shirt.

Subordinating conjunctions are the secondary class of conjunctions. They connect two unequal parts, one *main* (or *independent*) and the other *subordinate* (or *dependent*). I must go to the store *even though* I do not have enough money in the bank.

> *Because* I read the review, I do not want to go to the movie.

Notice that the presence of subordinating conjunctions makes clauses dependent. *I read the review* is an independent clause, but *because* makes the clause dependent. Thus, it needs an independent clause to complete the sentence.

Interjections

Interjections are words used to express emotion. Examples include *wow*, *ouch*, and *hooray*. Interjections are often separate from sentences; in those cases, the interjection is directly followed by an exclamation point. In other cases, the interjection is included in a sentence and followed by a comma. The punctuation plays a big role in the intensity of the emotion that the interjection is expressing. Using a comma or semicolon indicates less excitement than using an exclamation mark.

Capitalization Rules

Here's a non-exhaustive list of things that should be capitalized.

- the first word of every sentence
- the first word of every line of poetry
- the first letter of proper nouns (World War II)
- holidays (Valentine's Day)
- days of the week and months of the year (Tuesday, March)
- the first word, last word, and all major words in the titles of books, movies, songs, and other creative works (*To Kill a Mockingbird,* note that *a* is lowercase since it's not a major word, but *to* is capitalized since it's the first word of the title.
- titles when preceding a proper noun (President Roberto Gonzales, Aunt Judy)

When simply using a word such as president or secretary, though, the word is not capitalized.

> Officers of the new business must include a *president* and *treasurer*.

Seasons—spring, fall, etc.—are not capitalized.

North, *south*, *east*, and *west* are capitalized when referring to regions but are not when being used for directions. In general, if it's preceded by *the* it should be capitalized.

> I'm from the South.
> I drove south.

End Punctuation

Periods (.) are used to end a sentence that is a statement (*declarative*) or a command (*imperative*). They should not be used in a sentence that asks a question or is an exclamation. Periods are also used in abbreviations, which are shortened versions of words.

- Declarative: The boys refused to go to sleep.
- Imperative: Walk down to the bus stop.
- Abbreviations: Joan Roberts, M.D., Apple Inc., Mrs. Adamson
- If a sentence ends with an abbreviation, it is inappropriate to use two periods. It should end with a single period after the abbreviation.

The chef gathered the ingredients for the pie, which included apples, flour, sugar, etc.

Question marks (?) are used with direct questions (*interrogative*). An *indirect question* can use a period:

Interrogative: When does the next bus arrive?

Indirect Question: I wonder when the next bus arrives.

An *exclamation point (!)* is used to show strong emotion or can be used as an *interjection*. This punctuation should be used sparingly in formal writing situations.

What an amazing shot!

Whoa!

Commas

A *comma* (,) is the punctuation mark that signifies a pause—breath—between parts of a sentence. It denotes a break of flow. Proper comma usage helps readers understand the writer's intended emphasis of ideas.

In a complex sentence—one that contains a subordinate (dependent) clause or clauses—the use of a comma is dictated by where the subordinate clause is located. If the subordinate clause is located before the main clause, a comma is needed between the two clauses.

I will not pay for the steak, *because I don't have that much money*.

Generally, if the subordinate clause is placed after the main clause, no punctuation is needed.

I did well on my exam because I studied two hours the night before.

Notice how the last clause is dependent because it requires the earlier independent clauses to make sense.

Use a comma on both sides of an interrupting phrase.

I will pay for the ice cream, *chocolate and vanilla*, and then will eat it all myself.

The words forming the phrase in italics are nonessential (extra) information. To determine if a phrase is nonessential, try reading the sentence without the phrase and see if it's still coherent.

A comma is not necessary in this next sentence because no interruption—nonessential or extra information—has occurred. Read sentences aloud when uncertain.

I will pay for his chocolate and vanilla ice cream and then will eat it all myself.

If the nonessential phrase comes at the beginning of a sentence, a comma should only go at the end of the phrase. If the phrase comes at the end of a sentence, a comma should only go at the beginning of the phrase.

Other types of interruptions include the following:

- interjections: Oh no, I am not going.

- abbreviations: Barry Potter, M.D., specializes in heart disorders.
- direct addresses: Yes, Claudia, I am tired and going to bed.
- parenthetical phrases: His wife, lovely as she was, was not helpful.
- transitional phrases: Also, it is not possible.

The second comma in the following sentence is called an Oxford comma.

> I will pay for ice cream, syrup, and pop.

It is a comma used after the second-to-last item in a series of three or more items. It comes before the word *or* or *and*. Not everyone uses the Oxford comma; it is optional, but many believe it is needed. The comma functions as a tool to reduce confusion in writing. So, if omitting the Oxford comma would cause confusion, then it's best to include it.

Commas are used in math to mark the place of thousands in numerals, breaking them up so they are easier to read. Other uses for commas are in dates (*March 19, 2016*), letter greetings (*Dear Sally,*), and in between cities and states (*Louisville, KY*).

Semicolons

A *semicolon (;)* is used to connect ideas in a sentence in some way. There are three main ways to use semicolons.

Link two independent clauses without the use of a coordinating conjunction:

> I was late for work again; I'm definitely going to get fired.

Link two independent clauses with a transitional word:

> The songs were all easy to play; therefore, he didn't need to spend too much time practicing.

Between items in a series that are already separated by commas or if necessary to separate lengthy items in a list:

> Starbucks has locations in Media, PA; Swarthmore, PA; and Morton, PA.

> Several classroom management issues presented in the study: the advent of a poor teacher persona in the context of voice, dress, and style; teacher follow-through from the beginning of the school year to the end; and the depth of administrative support, including ISS and OSS protocol.

Colons

A *colon* is used after an independent clause to present an explanation or draw attention to what comes next in the sentence. There are several uses.

Explanations of ideas:

> They soon learned the hardest part about having a new baby: sleep deprivation.

Lists of items:

Shari picked up all the supplies she would need for the party: cups, plates, napkins, balloons, streamers, and party favors.

Time, subtitles, general salutations:

The time is 7:15.

I read a book entitled *Pluto: A Planet No More*.

To whom it may concern:

Parentheses and Dashes

Parentheses are half-round brackets that look like this: *( )*. They set off a word, phrase, or sentence that is an afterthought, explanation, or side note relevant to the surrounding text but not essential. A pair of commas is often used to set off this sort of information, but parentheses are generally used for information that would not fit well within a sentence or that the writer deems not important enough to be structurally part of the sentence.

The picture of the heart (see above) shows the major parts you should memorize.
Mount Everest is one of three mountains in the world that are over 28,000 feet high (K2 and Kanchenjunga are the other two).

See how the sentences above are complete without the parenthetical statements? In the first example, *see above* would not have fit well within the flow of the sentence. The second parenthetical statement could have been a separate sentence, but the writer deemed the information not pertinent to the topic.

The **em-dash** (—) is a mark longer than a hyphen used as a punctuation mark in sentences and to set apart a relevant thought. Even after plucking out the line separated by the dash marks, the sentence will be intact and make sense.

Looking out the airplane window at the landmarks—Lake Clarke, Thompson Community College, and the bridge—she couldn't help but feel excited to be home.

The dashes use is similar to that of parentheses or a pair of commas. So, what's the difference? Many believe that using dashes makes the clause within them stand out while using parentheses is subtler. It's advised to not use dashes when commas could be used instead.

Ellipses

An *ellipsis* (...) consists of three handy little dots that can speak volumes on behalf of irrelevant material. Writers use them in place of words, lines, phrases, list content, or paragraphs that might just as easily have been omitted from a passage of writing. This can be done to save space or to focus only on the specifically relevant material.

Exercise is good for some unexpected reasons. Watkins writes, "Exercise has many benefits such as...reducing cancer risk."

In the example above, the ellipsis takes the place of the other benefits of exercise that are more expected.

The ellipsis may also be used to show a pause in sentence flow.

"I'm wondering...how this could happen," Dylan said in a soft voice.

Quotation Marks

Double quotation marks are used at the beginning and end of a direct quote. They are also used with certain titles and to indicate that a term being used is slang or referenced in the sentence. Quotation marks should not be used with an indirect quote. Single quotation marks are used to indicate a quote within a quote.

Direct quote: "The weather is supposed to be beautiful this week," she said.

Indirect quote: One of the customers asked if the sale prices were still in effect.

Quote within a quote: "My little boy just said 'Mama, I want cookie,'" Maria shared.

Titles: Quotation marks should also be used to indicate titles of short works or sections of larger works, such as chapter titles. Other works that use quotation marks include poems, short stories, newspaper articles, magazine articles, web page titles, and songs.

"The Road Not Taken" is my favorite poem by Robert Frost.

"What a Wonderful World" is one of my favorite songs.

Specific or emphasized terms: Quotation marks can also be used to indicate a technical term or to set off a word that is being discussed in a sentence. Quotation marks can also indicate sarcasm.

The new step, called "levigation," is a very difficult technique.

He said he was "hungry" multiple times, but he only ate two bites.

Use with other punctuation: The use of quotation marks with other punctuation varies, depending on the role of the ending or separating punctuation.

In American English, *periods* and *commas* always go inside the quotation marks:

"This is the last time you are allowed to leave early," his boss stated.

The newscaster said, "We have some breaking news to report."

Question marks or *exclamation points* go inside the quotation marks when they are part of a direct quote:

The doctor shouted, "Get the crash cart!"

When the question mark or exclamation point is part of the sentence, not the quote, it should be placed outside of the quotation marks:

Was it Jackie that said, "Get some potatoes at the store"?

Apostrophes

This punctuation mark, the apostrophe (') is a versatile mark. It has several different functions:

- Quotes: Apostrophes are used when a second quote is needed within a quote.

    In my letter to my friend, I wrote, "The girl had to get a new purse, and guess what Mary did? She said, 'I'd like to go with you to the store.' I knew Mary would buy it for her."

- Contractions: Another use for an apostrophe in the quote above is a contraction. *I'd* is used for *I would.*

- Possession: An apostrophe followed by the letter s shows possession (Mary's purse). If the possessive word is plural, the apostrophe generally just follows the word. Not all possessive pronouns require apostrophes.

    The trees' leaves are all over the ground.

Hyphens

The *hyphen* (-) is a small hash mark that can be used to join words to show that they are linked.

Hyphenate two words that work together as a single adjective (a compound adjective).

honey-covered biscuits

Some words always require hyphens, even if not serving as an adjective.

merry-go-round

Hyphens always go after certain prefixes like *anti-* & *all-*.

Hyphens should also be used when the absence of the hyphen would cause a strange vowel combination (*semi-engineer*) or confusion. For example, *re-collect* should be used to describe something being gathered twice rather than being written as *recollect*, which means to remember.

Subjects

Every sentence must include a subject and a verb. The *subject* of a sentence is who or what the sentence is about. It's often directly stated and can be determined by asking "Who?" or "What?" did the action:

Most sentences contain a direct subject, in which the subject is mentioned in the sentence.

Kelly mowed the lawn.

Who mowed the lawn? *Kelly*

The air-conditioner ran all night

What ran all night? *the air-conditioner*

The subject of imperative sentences is the implied *you*, because imperative subjects are commands:

Go home after the meeting.

Who should go home after the meeting? *you* (implied)

In *expletive sentences* that start with "there are" or "there is," the subject is found after the predicate. The subject cannot be "there," so it must be another word in the sentence:

There is a cup sitting on the coffee table.

What is sitting on the coffee table? *a cup*

Simple and Complete Subjects

A *complete subject* includes the simple subject and all the words modifying it, including articles and adjectives. A *simple subject* is the single noun without its modifiers.

A warm, chocolate-chip cookie sat on the kitchen table.

Complete subject: *a warm, chocolate-chip cookie*

Simple subject: *cookie*

The words *a, warm, chocolate,* and *chip* all modify the simple subject *cookie.*

There might also be a *compound subject*, which would be two or more nouns without the modifiers.

A little girl and her mother walked into the shop.

Complete subject: *A little girl and her mother*

Compound subject: *girl, mother*

In this case, *the girl and her mother* are both completing the action of walking into the shop, so this is a *compound subject.*

Predicates

In addition to the subject, a sentence must also have a predicate. The *predicate* contains a verb and tells something about the subject. In addition to the verb, a predicate can also contain a direct or indirect object, object of a preposition, and other phrases.

The cats napped on the front porch.

In this sentence, cats is the subject because the sentence is about cats.

The *complete predicate* is everything else in the sentence: *napped on the front porch.* This phrase is the predicate because it tells us what the cats did.

This sentence can be broken down into a simple subject and predicate:

Cats napped.

In this sentence, *cats* is the simple subject, and *napped* is the *simple predicate*.

Although the sentence is very short and doesn't offer much information, it's still considered a complete sentence because it contains a subject and predicate.

Like a compound subject, a sentence can also have a ***compound predicate***. This is when the subject is or does two or more things in the sentence.

This easy chair reclines and swivels.

In this sentence, *this easy chair* is the complete subject. *Reclines and swivels* shows two actions of the chair, so this is the compound predicate.

Subject-Verb Agreement

The subject of a sentence and its verb must agree. The cornerstone rule of subject-verb agreement is that subject and verb must agree in number. Whether the subject is singular or plural, the verb must follow suit.

Incorrect: The houses is new.
Correct: The houses are new.
Also Correct: The house is new.

In other words, a singular subject requires a singular verb; a plural subject requires a plural verb. The words or phrases that come between the subject and verb do not alter this rule.

Incorrect: The houses built of brick is new.
Correct: The houses built of brick are new.

Incorrect: The houses with the sturdy porches is new.
Correct: The houses with the sturdy porches are new.

The subject will always follow the verb when a sentence begins with *here* or *there.* Identify these with care.

Incorrect: Here *is* the *houses* with sturdy porches.
Correct: Here *are* the *houses* with sturdy porches.

The subject in the sentences above is not *here*, it is *houses*. Remember, *here* and *there* are never subjects. Be careful that contractions such as *here's* or *there're* do not cause confusion!

Two subjects joined by *and* require a plural verb form, except when the two combine to make one thing:

Incorrect: Garrett and Jonathan is over there.
Correct: Garrett and Jonathan are over there.

Incorrect: Spaghetti and meatballs are a delicious meal!
Correct: Spaghetti and meatballs is a delicious meal!

In the example above, *spaghetti and meatballs* is a compound noun. However, *Garrett and Jonathan* is not a compound noun.

Two singular subjects joined by *or, either/or,* or *neither/nor* call for a singular verb form.

Incorrect: Butter or syrup are acceptable.
Correct: Butter or syrup is acceptable.

Plural subjects joined by *or*, *either/or*, or *neither/nor* are, indeed, plural.

The chairs or the boxes are being moved next.

If one subject is singular and the other is plural, the verb should agree with the closest noun.

Correct: The chair or the boxes are being moved next.
Correct: The chairs or the box is being moved next.

Some plurals of money, distance, and time call for a singular verb.

Incorrect: Three dollars *are* enough to buy that.
Correct: Three dollars *is* enough to buy that.

For words declaring degrees of quantity such as *many of, some of,* or *most of,* let the noun that follows of be the guide:

Incorrect: Many of the books is in the shelf.
Correct: Many of the books are in the shelf.

Incorrect: Most of the pie *are* on the table.
Correct: Most of the pie *is* on the table.

For indefinite pronouns like anybody or everybody, use singular verbs.

Everybody *is* going to the store.

However, the pronouns *few, many, several, all, some,* and *both* have their own rules and use plural forms.

Some *are* ready.

Some nouns like *crowd* and *congress* are called *collective nouns* and they require a singular verb form.

Congress *is* in session.
The news *is* over.

Books and movie titles, though, including plural nouns such as *Great Expectations*, also require a singular verb. Remember that only the subject affects the verb. While writing tricky subject-verb arrangements, say them aloud. Listen to them. Once the rules have been learned, one's ear will become sensitive to them, making it easier to pick out what's right and what's wrong.

Direct Objects

The *direct object* is the part of the sentence that receives the action of the verb. It is a noun and can usually be found after the verb. To find the direct object, first find the verb, and then ask the question *who* or *what* after it.

The bear climbed the tree.

What did the bear climb? *the tree*

Indirect Objects

An *indirect object* receives the direct object. It is usually found between the verb and the direct object. A strategy for identifying the indirect object is to find the verb and ask the questions *to whom/for whom* or *to what/ for what*.

Jane made her daughter a cake.

For whom did Jane make the cake? *her daughter*

Cake is the direct object because it is what Jane made, and *daughter* is the indirect object because she receives the cake.

Complements

A *complement* completes the meaning of an expression. A complement can be a pronoun, noun, or adjective. A verb complement refers to the direct object or indirect object in the sentence. An object complement gives more information about the direct object:

The magician got the kids excited.

Kids is the direct object, and *excited* is the object complement.

A *subject complement* comes after a linking verb. It is typically an adjective or noun that gives more information about the subject:

The king was noble and spared the thief's life.

Noble describes the *king* and follows the linking verb *was*.

Predicate Nouns

A *predicate noun* renames the subject:

John is a carpenter.

The subject is *John*, and the predicate noun is *carpenter.*

Predicate Adjectives

A *predicate adjective* describes the subject:

Margaret is beautiful.

The subject is *Margaret*, and the predicate adjective is *beautiful*.

Homonyms

Homonyms are words that sound the same but are spelled differently, and they have different meanings. There are several common homonyms that give writers trouble.

There, They're, and *Their*

The word *there* can be used as an adverb, adjective, or pronoun:

> *There* are ten children on the swim team this summer.

> I put my book over *there*, but now I can't find it.

The word *they're* is a contraction of the words *they* and *are*:

> *They're* flying in from Texas on Tuesday.

The word *their* is a possessive pronoun:

> I store *their* winter clothes in the attic.

Its and *It's*

Its is a possessive pronoun:

> The cat licked *its* injured paw.

It's is the contraction for the words *it* and *is*:

> *It's* unbelievable how many people opted not to vote in the last election.

Your and You're

Your is a possessive pronoun:

> Can I borrow *your* lawnmower this weekend?

You're is a contraction for the words *you* and *are*:

> *You're* about to embark on a fantastic journey.

To, Too, and *Two*

To is an adverb or a preposition used to show direction, relationship, or purpose:

> We are going *to* New York.

> They are going *to* see a show.

Too is an adverb that means more than enough, also, and very:

You have had *too* much candy.

We are on vacation that week, *too*.

Two is the written-out form of the numeral 2:

Two of the shirts didn't fit, so I will have to return them.

New and *Knew*

New is an adjective that means recent:

There's a *new* customer on the phone.

Knew is the past tense of the verb *know*:

I *knew* you'd have fun on this ride.

Affect and *Effect*

Affect and *effect* are complicated because they are used as both nouns and verbs, have similar meanings, and are pronounced the same.

	Affect	**Effect**
Noun Definition	emotional state	result
Noun Example	The patient's affect was flat.	The effects of smoking are well documented.
Verb Definition	to influence	to bring about
Verb Example	The pollen count affects my allergies.	The new candidate hopes to effect change.

Independent and Dependent Clauses

Independent and *dependent* clauses are strings of words that contain both a subject and a verb. An independent clause *can* stand alone as complete thought, but a dependent clause *cannot*. A dependent clause relies on other words to be a complete sentence.

Independent clause: The keys are on the counter.
Dependent clause: If the keys are on the counter

Notice that both clauses have a subject (*keys*) and a verb (*are*). The independent clause expresses a complete thought, but the word *if* at the beginning of the dependent clause makes it *dependent* on other words to be a complete thought.

Independent clause: If the keys are on the counter, please give them to me.

This presents a complete sentence since it includes at least one verb and one subject and is a complete thought. In this case, the independent clause has two subjects (*keys* & an implied *you*) and two verbs (*are* & *give*).

> Independent clause: I went to the store.
> Dependent clause: Because we are out of milk,
>
> Complete Sentence: Because we are out of milk, I went to the store.
> Complete Sentence: I went to the store because we are out of milk.

Phrases

A *phrase* is a group of words that do not make a complete thought or a clause. They are parts of sentences or clauses. Phrases can be used as nouns, adjectives, or adverbs. A phrase does not contain both a subject and a verb.

Prepositional Phrases

A *prepositional phrase* shows the relationship between a word in the sentence and the object of the preposition. The object of the preposition is a noun that follows the preposition.

> The orange pillows are on the couch.

On is the preposition, and *couch* is the object of the preposition.

> She brought her friend with the nice car.

With is the preposition, and *car* is the object of the preposition. Here are some common prepositions:

about	as	at	after
by	for	from	in
of	on	to	with

Verbals and Verbal Phrases

Verbals are forms of verbs that act as other parts of speech. They can be used as nouns, adjectives, or adverbs. Though they are verb forms, they are not to be used as the verb in the sentence. A word group that is based on a verbal is considered a *verbal phrase*. There are three major types of verbals: *participles, gerunds,* and *infinitives.*

Participles are verbals that act as adjectives. The present participle ends in *–ing*, and the past participle ends in *–d*, *-ed*, *-n*, or-*t*.

Verb	Present Participle	Past Participle
walk	walking	walked
share	sharing	shared

Participial phrases are made up of the participle and modifiers, complements, or objects.

Crying for most of an hour, the baby didn't seem to want to nap.

Having already taken this course, the student was bored during class.

Crying for most of an hour and *Having already taken this course* are the participial phrases.

Gerunds are verbals that are used as nouns and end in *–ing*. A gerund can be the subject or object of the sentence like a noun. Note that a present participle can also end in *–ing*, so it is important to distinguish between the two. The gerund is used as a noun, while the participle is used as an adjective.

Swimming is my favorite sport.

I wish I were sleeping.

A *gerund phrase* includes the gerund and any modifiers or complements, direct objects, indirect objects, or pronouns.

Cleaning the house is my least favorite weekend activity.

Cleaning the house is the gerund phrase acting as the subject of the sentence.

The most important goal this year is raising money for charity.

Raising money for charity is the gerund phrase acting as the direct object.

The police accused the woman of stealing the car.

The *gerund* phrase *stealing the car* is the object of the preposition in this sentence.

An *infinitive* is a verbal made up of the word to and a verb. Infinitives can be used as nouns, adjectives, or adverbs.

Examples: To eat, to jump, to swim, to lie, to call, to work

An *infinitive phrase* is made up of the infinitive plus any complements or modifiers. The infinitive phrase *to wait* is used as the subject in this sentence:

To wait was not what I had in mind.

The infinitive phrase *to sing* is used as the subject complement in this sentence:

Her dream is to sing.

The infinitive phrase *to grow* is used as an adverb in this sentence:

Children must eat to grow.

Appositive Phrases

An *appositive* is a noun or noun phrase that renames a noun that comes immediately before it in the sentence. An appositive can be a single word or several words. These phrases can be *essential* or *nonessential*. An essential appositive phrase is necessary to the meaning of the sentence and a nonessential appositive phrase is not. It is important to be able to distinguish these for purposes of comma use.

Essential: My sister Christina works at a school.

Naming which sister is essential to the meaning of the sentence, so no commas are needed.

Nonessential: My sister, who is a teacher, is coming over for dinner tonight.

Who is a teacher is not essential to the meaning of the sentence, so commas are required.

Absolute Phrases

An *absolute phrase* modifies a noun without using a conjunction. It is not the subject of the sentence and is not a complete thought on its own. Absolute phrases are set off from the independent clause with a comma.

Arms outstretched, she yelled at the sky.

All things considered, this has been a great day.

The Four Types of Sentence Structures

A *simple sentence* has one independent clause.

I am going to win.

A *compound sentence* has two independent clauses. A conjunction—*for, and, nor, but, or, yet, so*—links them together. Note that each of the independent clauses has a subject and a verb.

I am going to win, but the odds are against me.

A *complex sentence* has one independent clause and one or more dependent clauses.

I am going to win, even though I don't deserve it.

Even though I don't deserve it is a dependent clause. It does not stand on its own. Some conjunctions that link an independent and a dependent clause are *although*, *because*, *before*, *after*, *that*, *when*, *which*, and *while*.

A *compound-complex sentence* has at least three clauses, two of which are independent and at least one that is a dependent clause.

While trying to dance, I tripped over my partner's feet, but I regained my balance quickly.

The dependent clause is *While trying to dance*.

Sentence Fragments

A *sentence fragment* is an incomplete sentence. An independent clause is made up of a subject and a predicate, and both are needed to make a complete sentence.

Sentence fragments often begin with relative pronouns (when, which), subordinating conjunctions (because, although) or gerunds (trying, being, seeing). They might be missing the subject or the predicate.

The most common type of fragment is the isolated dependent clause, which can be corrected by joining it to the independent clause that appears before or after the fragment:

> Fragment: While the cookies baked.
>
> Correction: While the cookies baked, we played cards. (We played cards while the cookies baked.)

Run-on Sentences

A *run-on sentence* is created when two independent clauses (complete thoughts) are joined without correct punctuation or a conjunction. Run-on sentences can be corrected in the following ways:

- Join the independent clauses with a comma and coordinating conjunction.

  > Run-on: We forgot to return the library books we had to pay a fine.
  >
  > Correction: We forgot to return the library books, so we had to pay a fine.

- Join the independent clauses with a semicolon, dash, or colon when the clauses are closely related in meaning.

  > Run-on: I had a salad for lunch every day this week I feel healthier already.
  >
  > Correction: I had a salad for lunch every day this week; I feel healthier already.

- Join the independent clauses with a *semicolon and a conjunctive adverb.*

  > Run-on: We arrived at the animal shelter on time however the dog had already been adopted.
  >
  > Correction: We arrived at the animal shelter on time; however, the dog had already been adopted.

- Separate the independent clauses into two sentences *with a period.*

  > Run-on: He tapes his favorite television show he never misses an episode.
  >
  > Correction: He tapes his favorite television show. He never misses an episode.

- *Rearrange the wording* of the sentence to create an independent clause and a dependent clause.

    Run-on: My wedding date is coming up I am getting more excited to walk down the aisle.

    Correction: As my wedding date approaches, I am getting more excited to walk down the aisle.

Dangling and Misplaced Modifiers

A *modifier* is a phrase that describes, alters, limits, or gives more information about a word in the sentence. The two most common issues are dangling and misplaced modifiers.

A *dangling modifier* is created when the phrase modifies a word that is not clearly stated in the sentence.

Dangling modifier: Having finished dinner, the dishes were cleared from the table.

Correction: Having finished dinner, Amy cleared the dishes from the table.

In the first sentence, *having finished dinner* appears to modify *the dishes*, which obviously can't finish dinner. The second sentence adds the subject *Amy*, to make it clear who has finished dinner.

Dangling modifier: Hoping to improve test scores, all new books were ordered for the school.

Correction: Hoping to improve test scores, administrators ordered all new books for the school.

Without the subject *administrators*, it appears the books are hoping to improve test scores, which doesn't make sense.

Misplaced modifiers are placed incorrectly in the sentence, which can cause confusion. Compare these examples:

Misplaced modifier: Rory purchased a new flat screen television and placed it on the wall above the fireplace, with all the bells and whistles.

Revised: Rory purchased a new flat screen television, with all the bells and whistles, and placed it on the wall above the fireplace.

The bells and whistles should modify the television, not the fireplace.

Misplaced modifier: The delivery driver arrived late with the pizza, who was usually on time.

Revised: The delivery driver, who usually was on time, arrived late with the pizza.

This suggests that the delivery driver was usually on time, instead of the pizza.

Misplaced modifier: We saw a family of ducks on the way to church.

Revised: On the way to church, we saw a family of ducks.

The misplaced modifier, here, suggests the *ducks* were on their way to church, instead of the pronoun *we*.

Split Infinitives

An infinitive is made up of the word *to* and a verb, such as: to run, to jump, to ask. A *split infinitive* is created when a word comes between *to* and the verb.

Split infinitive: To quickly run

Correction: To run quickly

Split infinitive: To quietly ask

Correction: To ask quietly

Double Negatives

A *double negative* is a negative statement that includes two negative elements. This is incorrect in Standard English.

Incorrect: She hasn't never come to my house to visit.

Correct: She has never come to my house to visit.

The intended meaning is that she has never come to the house, so the double negative is incorrect. However, it is possible to use two negatives to create a positive statement.

Correct: She was not unhappy with her performance on the quiz.

In this case, the double negative, *was not unhappy*, is intended to show a positive, so it is correct. This means that she was somewhat happy with her performance.

Faulty Parallelism

It is necessary to use parallel construction in sentences that have multiple similar ideas. Using parallel structure provides clarity in writing. *Faulty parallelism* is created when multiple ideas are joined using different sentence structures. Compare these examples:

Incorrect: We start each practice with stretches, a run, and fielding grounders.
Correct: We start each practice with stretching, running, and fielding grounders.

Incorrect: I watched some television, reading my book, and fell asleep.
Correct: I watched some television, read my book, and fell asleep.

Incorrect: *Some of the readiness skills for Kindergarten are to cut with scissors, to tie shoes, and dressing independently.*
Correct: *Some of the readiness skills for Kindergarten are being able to cut with scissors, to tie shoes, and to dress independently.*

Subordination

If multiple pieces of information in a sentence are not equal, they can be joined by creating an independent clause and a dependent clause. The less important information becomes the *subordinate clause*:

Draft: The hotel was acceptable. We wouldn't stay at the hotel again.

Revised: Though the hotel was acceptable, we wouldn't stay there again.

The more important information (*we wouldn't stay there again*) becomes the main clause, and the less important information (*the hotel was acceptable*) becomes the subordinate clause.

Context Clues

Context clues help readers understand unfamiliar words, and thankfully, there are many types.

Synonyms are words or phrases that have nearly, if not exactly, the same meaning as other words or phrases

Large boxes are needed to pack *big* items.

Antonyms are words or phrases that have opposite definitions. Antonyms, like synonyms, can serve as context clues, although more cryptically.

Large boxes are not needed to pack *small* items.

Definitions are sometimes included within a sentence to define uncommon words.

They practiced the *rumba*, a *type of dance*, for hours on end.

Explanations provide context through elaboration.

Large boxes holding items weighing over 60 pounds were stacked in the corner.

Here's an example of *contrast*:

These *minute* creatures were much different than the *huge* mammals that the zoologist was accustomed to dealing with.

Beware of Simplicity

Sometimes the answer may seem very simple. In this case, it's prudent to look more carefully at the question and the possible answer choices. Very brief answers aren't always correct, and the opposite may also be true. The goal is to read all the answer choices carefully, trying to rule out those that don't make sense.

Final Notes

It's best to read every answer choice before making a decision. While some answers may seem plausible, there may be others that are better choices. First instinct is usually right, but reading every answer is recommended. Caution should be taken in choosing an answer that "sounds right." Grammar rules can be tricky, and what sounds right may not be correct. It's best to rely on knowledge of grammar to choose the best answer. Ruling out incorrect responses can help narrow the choices down. Choosing between two choices (after reading them carefully) and selecting the answer that best matches the rules of Standard English is less overwhelming.

Practice Questions

Sentence Correction

Directions for questions 1–10

Select the best version of the underlined part of the sentence. The first choice is the same as the original sentence. If you think the original sentence is best, choose the first answer.

1. An important issues stemming from this meeting is that we won't have enough time to meet all of the objectives.
   a. An important issues stemming from this meeting
   b. Important issue stemming from this meeting
   c. An important issue stemming from this meeting
   d. Important issues stemming from this meeting

2. The rising popularity of the clean eating movement can be attributed to the fact that experts say added sugars and chemicals in our food are to blame for the obesity epidemic.
   a. to the fact that experts say added sugars and chemicals in our food are to blame for the obesity epidemic.
   b. in the facts that experts say added sugars and chemicals in our food are to blame for the obesity epidemic.
   c. to the fact that experts saying added sugars and chemicals in our food are to blame for the obesity epidemic.
   d. with the facts that experts say added sugars and chemicals in our food are to blame for the obesity epidemic.

3. She's looking for a suitcase that can fit all of her clothes, shoes, accessory, and makeup.
   a. clothes, shoes, accessory, and makeup.
   b. clothes, shoes, accessories, and makeup.
   c. clothes, shoes, accessories, and makeups.
   d. clothes, shoe, accessory, and makeup.

4. Shawn started taking guitar lessons while he wanted to become a better musician.
   a. while he wanted to become a better musician.
   b. because he wants to become a better musician.
   c. even though he wanted to become a better musician.
   d. because he wanted to become a better musician.

5. Considering the recent rains we have had, it's a wonder the plants haven't drowned.
   a. Considering the recent rains we have had, it's a wonder
   b. Consider the recent rains we have had, it's a wonder
   c. Considering for how much recent rain we have had, its a wonder
   d. Considering, the recent rains we have had, its a wonder

6. Since none of the furniture were delivered on time, we have to move in at a later date.
    a. Since none of the furniture were delivered on time,
    b. Since none of the furniture was delivered on time,
    c. Since all of the furniture were delivered on time,
    d. Since all of the furniture was delivered on time

7. It is necessary for instructors to offer tutoring to any students who need extra help in the class.
    a. to any students who need extra help in the class.
    b. for any students that need extra help in the class.
    c. with any students who need extra help in the class.
    d. for any students needing any extra help in their class.

8. The fact the train set only includes four cars and one small track was a big disappointment to my son.
    a. the train set only includes four cars and one small track was a big disappointment
    b. that the trains set only include four cars and one small track was a big disappointment
    c. that the train set only includes four cars and one small track was a big disappointment
    d. that the train set only includes four cars and one small track were a big disappointment

9. Because many people feel there are too many distractions to get any work done, I actually enjoy working from home.
    a. Because many people
    b. While many people
    c. Maybe many people
    d. With most people

10. There were many questions about what causes the case to have gone cold, but the detective wasn't willing to discuss it with reporters.
    a. about what causes the case to have gone cold
    b. about why the case is cold
    c. about what causes the case to go cold
    d. about why the case went cold

Construction Shift

Directions for questions 11–20

Rewrite the sentence in your head following the directions given below. Keep in mind that your new sentence should be well written and should have essentially the same meaning as the original sentence.

11. Although she was nervous speaking in front of a crowd, the author read her narrative with poise and confidence.

Rewrite, beginning with

The author had poise and confidence while reading

The next words will be

a. because she was nervous speaking in front of a crowd.
b. but she was nervous speaking in front of a crowd.
c. even though she was nervous speaking in front of a crowd.
d. before she was nervous speaking in front of a crowd.

12. There was a storm surge and loss of electricity during the hurricane.

Rewrite, beginning with

While the hurricane occurred,

The next words will be

a. there was a storm surge after the electricity went out.
b. the storm surge caused the electricity to go out.
c. the electricity surged into the storm.
d. the electricity went out, and there was a storm surge.

13. When one elephant in a herd is sick, the rest of the herd will help it walk and bring it food.

Rewrite, beginning with

An elephant herd will

The next words will be

a. be too sick and tired to walk
b. help and support
c. gather food when they're sick
d. be unable to walk without food

14. They went out to eat after the soccer game.

Rewrite, beginning with

They finished the soccer game

The next words will be

a. then went out to eat.
b. after they went out to eat.
c. so they could go out to eat.
d. because they went out to eat.

15. Armani got lost when she walked around Paris.

Rewrite, beginning with

Walking through Paris,

The next words will be
a. you can get lost.
b. Armani found herself lost.
c. she should have gotten lost.
d. is about getting lost.

16. After his cat died, Phoenix buried the cat with her favorite toys in his backyard.

Rewrite, beginning with

Phoenix buried his cat

The next words will be
a. in his backyard before she died.
b. after she died in the backyard.
c. with her favorite toys after she died.
d. after he buried her toys in the backyard.

17. While I was in the helicopter, I saw the sunset, and tears streamed down my eyes.

Rewrite, beginning with

Tears streamed down my eyes

The next words will be:
a. while I watched the helicopter fly into the sunset.
b. because the sunset flew up into the sky.
c. because the helicopter was facing the sunset.
d. when I saw the sunset from the helicopter.

18. I won't go to the party unless some of my friends go.

Rewrite, beginning with

I will go the party

The next words will be
a. if I want to.
b. if my friends go.
c. since a couple of my friends are going.
d. unless people I know go.

19. He had a broken leg before the car accident, so it took him a long time to recover.

Rewrite, beginning with

He took a long time to recover from the car accident

The next words will be

a. from his two broken legs.
b. after he broke his leg.
c. because he already had a broken leg.
d. since he broke his leg again afterward.

20. We had a party the day after Halloween to celebrate my birthday.

Rewrite, beginning with

It was my birthday.

The next words will be

a. , so we celebrated with a party the day after Halloween.
b. the day of Halloween so we celebrated with a party.
c. , and we celebrated with a Halloween party the day after.
d. a few days before Halloween, so we threw a party.

Answer Explanations

Sentence Correction

1. C: In this answer, the article and subject agree, and the subject and predicate agree. Choice *A* is incorrect because the article (*an*) and the noun (*issues*) do not agree in number. Choice *B* is incorrect because an article is needed before *important issue*. Choice *D* is incorrect because the plural subject *issues* does not agree with the singular verb *is*.

2. A: Choices *B* and *D* both use the expression *attributed to the fact* incorrectly. It can only be attributed *to* the fact, not *with* or *in* the fact. Choice *C* incorrectly uses a gerund, *saying*, when it should use the present tense of the verb *say*.

3. B: Choice *B* is correct because it uses correct parallel structure of plural nouns. *A* is incorrect because the word *accessory* is in singular form. Choice *C* is incorrect because it pluralizes *makeup*, which is already in plural form. Choice *D* is incorrect because it again uses the singular *accessory*, and it uses the singular *shoe*.

4: In a cause/effect relationship, it is correct to use the word because in the clausal part of the sentence. This can eliminate both Choices *A and* C which don't clearly show the cause/effect relationship. Choice *B* is incorrect because it uses the present tense, when the first part of the sentence is in the past tense. It makes grammatical sense for both parts of the sentence to be in present tense.

5. A: In Choice *B*, the present tense form of the verb *consider* creates an independent clause joined to another independent clause with only a comma, which is a comma splice and grammatically incorrect. Both *C* and *D* use the possessive form of *its*, when it should be the contraction *it's* for *it is*. Choice *D* also includes incorrect comma placement.

6. B: Choice *A* uses the plural form of the verb, when the subject is the pronoun *none*, which needs a singular verb. Choice *C* also uses the wrong verb form and uses the word *all* in place of *none*, which doesn't make sense in the context of the sentence. Choice *D* uses *all* again, and is missing the comma, which is necessary to set the dependent clause off from the independent clause.

7. A: Answer Choice *A* uses the best, most concise word choice. Choice *B* uses the pronoun *that* to refer to people instead of *who*. *C* incorrectly uses the preposition *with*. Choice *D* uses the preposition *for* and the additional word *any*, making the sentence wordy and less clear.

8. C: Choice *A* is missing the word *that*, which is necessary for the sentence to make sense. Choice *B* pluralizes *trains* and uses the singular form of the word *include*, so it does not agree with the word *set*. Choice *D* changes the verb to *were*, which is in plural form and does not agree with the singular subject.

9. B: Choice *B* uses the best choice of words to create a subordinate and independent clause. In Choice *A*, *because* makes it seem like this is the reason I enjoy working from home, which is incorrect. In *C*, the word *maybe* creates two independent clauses, which are not joined properly with a comma. Choice *D* uses *with*, which does not make grammatical sense.

10. D: Choices *A* and *C* use additional words and phrases that are not necessary. Choice *B* is more concise, but uses the present tense of *is*. This does not agree with the rest of the sentence, which uses past tense. The best choice is Choice *D*, which uses the most concise sentence structure and is grammatically correct.

Construction Shift

11. C: The original sentence states that despite the author being nervous, she was able to read with poise and confidence, which is stated in Choice *C*. Choice *A* changes the meaning by adding *because*; however, the author didn't read with confidence *because* she was nervous, but *despite* being nervous. Choice *B* is closer to the original meaning; however, it loses the emphasis of her succeeding *despite* her condition. Choice *D* adds the word *before*, which doesn't make much sense on its own, much less in relation to the original sentence.

12. D: The original sentence states that there was a storm surge and loss of electricity during the hurricane, making Choice *D* correct. Choices *A* and *B* arrange the storm surge and the loss of electricity within a cause and effect statement, which changes the meaning of the original sentence. Choice *C* changes *surge* from a noun into a verb and creates an entirely different situation.

13. B: The original sentence states that an elephant herd will help and support another herd member if it is sick, so Choice *B* is correct. Choice *A* is incorrect because it states the whole herd will be too sick and too tired to walk instead of a single elephant. Choice *C* is incorrect because the original sentence does not say that the herd gathers food when *they* are sick, but when a single member of the herd is sick. Although Choice *D* might be correct in a general sense, it does not relate to the meaning of the original sentence and is therefore incorrect.

14. A: The original sentence says that after a soccer game, they went out to eat. Choice *A* shows the same sequence: they finished the soccer game *then* went out to eat. Choice *B* is incorrect because it reverses the sequence of events. Choices *C* and *D* are incorrect because the words *so* and *because* change the meaning of the original sentence.

15. B: Choice *B* is correct because the idea of the original sentences is Armani getting lost while walking through Paris. Choice *A* is incorrect because it replaces third person with second person. Choice *C* is incorrect because the word *should* indicates an obligation to get lost. Choice *D* is incorrect because it is not specific to the original sentence but instead makes a generalization about getting lost.

16. C: Choice *C* is correct because it shows that Phoenix buried his cat with her favorite toys after she died, which is true of the original statement. Although Choices *A*, *B*, and *D* mention a backyard, the meanings of these choices are skewed. Choice *A* says that Phoenix buried his cat alive, which is incorrect. Choice *B* says his cat died in the backyard, which we do not know to be true. Choice *D* says Phoenix buried his cat after he buried her toys, which is also incorrect.

17. D: Choice *D* is correct because it expresses the sentiment of a moment of joy bringing tears to one's eyes as one sees a sunset while in a helicopter. Choice *A* is incorrect because it implies that the person was outside of the helicopter watching it from afar. Choice *B* is incorrect because the original sentence does not portray the sunset *flying up* into the sky. Choice *C* is incorrect because, while the helicopter may have been facing the sunset, this is not the reason that tears were in the speaker's eyes.

18. B: *B* is correct because like the original sentence, it expresses their plan to go to the party if friends also go. Choice *A* is incorrect because it does not follow the meaning of the original sentence. Choice *C* is incorrect because it states that their friends are going, even though that is not known. Choice *D* is incorrect because it would make the new sentence mean the opposite of the original sentence.

19. C: Choice *C* is correct because the original sentence states that his recovery time was long because his leg was broken before the accident. Choice *A* is incorrect because there is no indication that the man had two broken legs. Choice *B* is incorrect because it indicates that he broke his leg during the car accident, not before. Choice *D* is incorrect because there is no indication that he broke his leg after the car accident.

20. A: Choice *A* is correct because it expresses the fact that the birthday and the party were both after Halloween. Choice *B* is incorrect because it says that the birthday was on Halloween, even though that was not stated in the original sentence. Choice *C* is incorrect because it says the party was specifically a Halloween party and not a birthday party. Choice *D* is incorrect because the party was after Halloween, not before.

Essay Revision

Revisions

Leaving a few minutes at the end to revise and proofread offers an opportunity for writers to polish things up. Putting one's self in the reader's shoes and focusing on what the essay actually says helps writers identify problems—it's a movement from the mindset of writer to the mindset of editor. The goal is to have a clean, clear copy of the essay. The following areas should be considered when proofreading:

- Sentence fragments
- Awkward sentence structure
- Run-on sentences
- Incorrect word choice
- Grammatical agreement errors
- Spelling errors
- Punctuation errors
- Capitalization errors

Organization

Good writing is not merely a random collection of sentences. No matter how well written, sentences must relate and coordinate appropriately with one another. If not, the writing seems random, haphazard, and disorganized. Therefore, good writing must be organized, where each sentence fits a larger context and relates to the sentences around it.

Transition Words

The writer should act as a guide, showing the reader how all the sentences fit together. Consider the seat belt example again:

> Seat belts save more lives than any other automobile safety feature. Many studies show that airbags save lives as well. Not all cars have airbags. Many older cars don't. Air bags aren't entirely reliable. Studies show that in 15% of accidents, airbags don't deploy as designed. Seat belt malfunctions are extremely rare.

There's nothing wrong with any of these sentences individually, but together they're disjointed and difficult to follow. The best way for the writer to communicate information is through the use of transition words. Here are examples of transition words and phrases that tie sentences together, enabling a more natural flow:

- To show causality: *as a result*, *therefore*, and *consequently*
- To compare and contrast: *however*, *but*, and *on the other hand*
- To introduce examples: *for instance*, *namely*, and *including*
- To show order of importance: *foremost*, *primarily*, *secondly*, and *lastly*

NOTE: This is not a complete list of transitions. There are many more that can be used; however, most fit into these or similar categories. The important point is that the words should clearly show the relationship between sentences, supporting information, and the main idea.

Here is an update to the previous example using transition words. These changes make it easier to read and bring clarity to the writer's points:

> Seat belts save more lives than any other automobile safety feature. Many studies show that airbags save lives as well; however, not all cars have airbags. For instance, some older cars don't. Furthermore, air bags aren't entirely reliable. For example, studies show that in 15% of accidents, airbags don't deploy as designed, but, on the other hand, seat belt malfunctions are extremely rare.

Also, be prepared to analyze whether the writer is using the best transition word or phrase for the situation. Take this sentence for example: "As a result, seat belt malfunctions are extremely rare." This sentence doesn't make sense in the context above because the writer is trying to show the contrast between seat belts and airbags, not the causality.

Logical Sequence

Even if the writer includes plenty of information to support their point, the writing is only coherent when the information is in a logical order. First, the writer should introduce the main idea, whether for a paragraph, a section, or the entire piece. Second, they should present evidence to support the main idea by using transitional language. This shows the reader how the information relates to the main idea and to the sentences around it. The writer should then take time to interpret the information, making sure necessary connections are obvious to the reader. Finally, the writer can summarize the information in a closing section.

Though most writing follows this pattern, it isn't a set rule. Sometimes writers change the order for effect. For example, the writer can begin with a surprising piece of supporting information to grab the reader's attention, and then transition to the main idea. Thus, if a passage doesn't follow the logical order, don't immediately assume it's wrong. However, most writing usually settles into a logical sequence after a nontraditional beginning.

Introductions and Conclusions

Examining the writer's strategies for introductions and conclusions puts the reader in the right mindset to interpret the rest of the text. Look for methods the writer might use for introductions such as:

- Stating the main point immediately, followed by outlining how the rest of the piece supports this claim.
- Establishing important, smaller pieces of the main idea first, and then grouping these points into a case for the main idea.
- Opening with a quotation, anecdote, question, seeming paradox, or other piece of interesting information, and then using it to lead to the main point.

Whatever method the writer chooses, the introduction should make their intention clear, establish their voice as a credible one, and encourage a person to continue reading.

Conclusions tend to follow a similar pattern. In them, the writer restates their main idea a final time, often after summarizing the smaller pieces of that idea. If the introduction uses a quote or anecdote to grab the reader's attention, the conclusion often makes reference to it again. Whatever way the writer chooses to arrange the conclusion, the final restatement of the main idea should be clear and simple for the reader to interpret. Finally, conclusions shouldn't introduce any new information.

Precision

People often think of precision in terms of math, but precise word choice is another key to successful writing. Since language itself is imprecise, it's important for the writer to find the exact word or words to convey the full, intended meaning of a given situation. For example:

> The number of deaths has gone down since seat belt laws started.

There are several problems with this sentence. First, the word *deaths* is too general. From the context, it's assumed that the writer is referring only to deaths caused by car accidents. However, without clarification, the sentence lacks impact and is probably untrue. The phrase "gone down" might be accurate, but a more precise word could provide more information and greater accuracy. Did the numbers show a slow and steady decrease of highway fatalities or a sudden drop? If the latter is true, the writer is missing a chance to make their point more dramatically. Instead of "gone down" they could substitute *plummeted*, *fallen drastically*, or *rapidly diminished* to bring the information to life. Also, the phrase "seat belt laws" is unclear. Does it refer to laws requiring cars to include seat belts or to laws requiring drivers and passengers to use them? Finally, *started* is not a strong verb. Words like *enacted* or *adopted* are more direct and make the content more real. When put together, these changes create a far more powerful sentence:

> The number of highway fatalities has plummeted since laws requiring seat belt usage were enacted.

However, it's important to note that precise word choice can sometimes be taken too far. If the writer of the sentence above takes precision to an extreme, it might result in the following:

The incidence of high-speed, automobile accident related fatalities has decreased 75% and continued to remain at historical lows since the initial set of federal legislations requiring seat belt use were enacted in 1992.

This sentence is extremely precise, but it takes so long to achieve that precision that it suffers from a lack of clarity. Precise writing is about finding the right balance between information and flow. This is also an issue of conciseness (discussed in the next section).

The last thing to consider with precision is a word choice that's not only unclear or uninteresting, but also confusing or misleading. For example:

The number of highway fatalities has become hugely lower since laws requiring seat belt use were enacted.

In this case, the reader might be confused by the word *hugely*. Huge means large, but here the writer uses *hugely* to describe something small. Though most readers can decipher this, doing so disconnects them from the flow of the writing and makes the writer's point less effective.

Conciseness

"Less is more" is a good rule to follow when writing a sentence. Unfortunately, writers often include extra words and phrases that seem necessary at the time but add nothing to the main idea. This confuses the reader and creates unnecessary repetition. Writing that lacks conciseness is usually guilty of excessive wordiness and redundant phrases. Here's an example containing both of these issues:

> When legislators decided to begin creating legislation making it mandatory for automobile drivers and passengers to make use of seat belts while in cars, a large number of them made those laws for reasons that were political reasons.

There are several empty or "fluff" words here that take up too much space. These can be eliminated while still maintaining the writer's meaning. For example:

- "Decided to begin" could be shortened to "began"
- "Making it mandatory for" could be shortened to "requiring"
- "Make use of" could be shortened to "use"
- "A large number" could be shortened to "many"

In addition, there are several examples of redundancy that can be eliminated:

- "Legislators decided to begin creating legislation" and "made those laws"
- "Automobile drivers and passengers" and "while in cars"
- "Reasons that were political reasons"

These changes are incorporated as follows:

> When legislators began requiring drivers and passengers to use seat belts, many of them did so for political reasons.

There are many general examples of redundant phrases, such as "add an additional," "complete and total," "time schedule," and "transportation vehicle." If asked to identify a redundant phrase on the test, look for words that are close together with the same (or similar) meanings.

Agreement

Agreement in Number

Subjects and verbs must agree in number. If a sentence has a singular subject, then it must use a singular verb. If there is a plural subject, then it must use a plural verb.

Singular Noun	Singular Verb	Plural Noun	Plural verb
Man	has	men	have
child	plays	children	play
basketball	bounces	basketballs	bounce

Agreement in Person

Verbs must also agree in person. A subject that uses first person must include a verb that is also in first person. The same is true for second and third person subjects and verbs.

	Noun	**Verb**
First person	I	am
Second person	you	are
Third person	she	is

Common Agreement Errors

Compound Subjects

Compound subjects are when two or more subjects are joined by a coordinating conjunction, such as *and, or, neither*, or *nor*. Errors in agreement sometimes occur when there is a compound subject:

Incorrect: Mike and I am in a meeting this morning.

Correct: Mike and I are in a meeting this morning.

A compound subject always uses the plural form of the verb to match with the plural subject. In the above example, readers can substitute "Mike and I" for "we" to make it easier to determine the verb: "We *are* in a meeting this morning."

Separation of Subject and Verb

Errors sometimes occur when the subject is separated from the verb by a prepositional phrase or parenthetical element:

Incorrect: The objective of the teachers are to help students learn.

Correct: The objective of the teachers is to help students learn.

The verb must agree with the singular subject *objective*, not the word *teachers*, which is the object of the preposition *of* and does not influence the subject. An easy way to determine if the subject and verb agree is to take out the middle preposition: "The objective *is* to help students learn."

Indefinite Pronouns

Indefinite pronouns refer to people or groups in a general way: *each, anyone, none, all, either, neither,* and *everyone.* Some indefinite pronouns are always singular, such as *each, everyone, someone*, and *everybody*, which affects verb choice:

Incorrect: Each of them are competing in the race.

Correct: Each of them is competing in the race.

While the word *them* can indicate that a plural verb is needed, the subject *each* is singular regardless of what it refers to, requiring the singular verb, *is*.

Other indefinite pronouns can be singular or plural, depending on what they are referring to, such as *anyone, all,* and *some.*

Some of the orders are scheduled to arrive today.

Some refers to *orders*, which is plural, so the plural verb (*are*) is needed.

Some of the cake is left on the dining room table.

Some refers to *cake*, which is singular, so the singular verb (*is*) is needed.

Subjects Joined by Or and Nor

Compound subjects joined by *or* or *nor* rely on the subject nearest to the verb to determine conjugation and agreement:

Neither Ben nor Jeff was in attendance at the conference.

Pink or purple is the bride's color choice.

In each example, the subjects are both singular, so the verb should be singular.

If one subject is singular and the other plural, the subject nearest to the verb is the one that needs to agree:

Either the shirt or pants are hanging on the clothesline.

In this example there is a singular subject (*shirt*) and a plural subject (*pants*), so the verb (*are*) should agree with the subject nearest to it (*pants*).

Collective Nouns

Collective nouns can use a singular or plural verb depending on their function in the sentence. If the collective noun is acting as a unit, then a singular verb is needed. Otherwise, it's necessary to use a plural verb.

The staff is required to meet every third Friday of the month.

The *staff* is meeting as a collective unit, so a singular verb is needed.

The staff are getting in their cars to go home.

The staff get into their cars separately, so a plural verb is needed.

Plural Nouns with Singular Meaning

Certain nouns end in *s*, like a plural noun, but have singular meaning, such as *mathematics, news,* and *civics*. These nouns should use a singular verb.

The news is on at 8:00 tonight.

Nouns that are single things, but have two parts, are considered plural and should use a plural verb, such as *scissors, pants,* and *tweezers.*

My favorite pants are in the washing machine.

There Is and There Are

There cannot be a subject, so verb agreement should be based on a word that comes after the verb.

There is a hole in the road.

The subject in this sentence is *hole*, which is singular, so the verb should be singular (*is*).

There are kids playing kickball in the street.

The subject in this sentence is *kids*, which is plural, so the verb should be plural (*are*).

Verb Tense

Shifting verb forms entails conjugation, which is used to indicate tense, voice, or mood.

Verb tense is used to show when the action in the sentence took place. There are several different verb tenses, and it is important to know how and when to use them. Some verb tenses can be achieved by changing the form of the verb, while others require the use of helping verbs (e.g., *is, was,* or *has*).

- *Present tense* shows the action is happening currently or is ongoing:

  I walk to work every morning.

  She is stressed about the deadline.

- *Past tense* shows that the action happened in the past or that the state of being is in the past:

  I walked to work yesterday morning.

  She was stressed about the deadline.

- *Future tense* shows that the action will happen in the future or is a future state of being:

  I will walk to work tomorrow morning.

  She will be stressed about the deadline.

- *Present perfect tense* shows action that began in the past, but continues into the present:

  I have walked to work all week.

  She has been stressed about the deadline.

- *Past perfect tense* shows an action was finished before another took place:

  I had walked all week until I sprained my ankle.

  She had been stressed about the deadline until we talked about it.

- *Future perfect tense* shows an action that will be completed at some point in the future:

  By the time the bus arrives, I will have walked to work already.

Sentence Structure

Sentence Types

There are four ways in which we can structure sentences: simple, compound, complex, and compound-complex. Sentences can be composed of just one clause or many clauses joined together.

When a sentence is composed of just one clause (an independent clause), we call it a simple sentence. Simple sentences do not necessarily have to be short sentences. They just require one independent clause with a subject and a predicate. For example:

Thomas marched over to Andrew's house.

Jonah and Mary constructed a simplified version of the Eiffel Tower with Legos.

When a sentence has two or more independent clauses we call it a compound sentence. The clauses are connected by a comma and a coordinating conjunction—*and, but, or, nor, for*—or by a semicolon. Compound sentences do not have dependent clauses. For example:

We went to the fireworks stand, and we bought enough fireworks to last all night.

The children sat on the grass, and then we lit the fireworks one at a time.

When a sentence has just one independent clause and includes one or more dependent clauses, we call it a complex sentence:

Because she slept well and drank coffee, Sarah was quite productive at work.

Although Will had coffee, he made mistakes while using the photocopier.

When a sentence has two or more independent clauses and at least one dependent clause, we call it a compound-complex sentence:

It may come as a surprise, but I found the tickets, and you can go to the show.

Jade is the girl who dove from the high-dive, and she stunned the audience silent.

Sentence Fragments

Remember that a complete sentence must have both a subject and a verb. Complete sentences consist of at least one independent clause. Incomplete sentences are called sentence fragments. A sentence fragment is a common error in writing. Sentence fragments can be independent clauses that start with subordinating words, such as *but, as, so that,* or *because,* or they could simply be missing a subject or verb.

You can correct a fragment error by adding the fragment to a nearby sentence or by adding or removing words to make it an independent clause. For example:

> Dogs are my favorite animals. Because cats are too independent. (Incorrect; the word because creates a sentence fragment)
>
> Dogs are my favorite animals because cats are too independent. (Correct; the fragment becomes a dependent clause.)
>
> Dogs are my favorite animals. Cats are too independent. (Correct; the fragment becomes a simple sentence.)

Run-on Sentences

Another common mistake in writing is the run-on sentence. A run-on is created when two or more independent clauses are joined without the use of a conjunction, a semicolon, a colon, or a dash. We don't want to use commas where periods belong. Here is an example of a run-on sentence:

> Making wedding cakes can take many hours I am very impatient, I want to see them completed right away.

There are a variety of ways to correct a run-on sentence. The method you choose will depend on the context of the sentence and how it fits with neighboring sentences:

> Making wedding cakes can take many hours. I am very impatient. I want to see them completed right away. (Use periods to create more than one sentence.)
>
> Making wedding cakes can take many hours; I am very impatient—I want to see them completed right away. (Correct the sentence using a semicolon, colon, or dash.)
>
> Making wedding cakes can take many hours and I am very impatient, so I want to see them completed right away. (Correct the sentence using coordinating conjunctions.)
>
> I am very impatient because I would rather see completed wedding cakes right away than wait for it to take many hours. (Correct the sentence by revising.)

Dangling and Misplaced Modifiers

A modifier is a word or phrase meant to describe or clarify another word in the sentence. When a sentence has a modifier but is missing the word it describes or clarifies, it's an error called a dangling modifier. We can fix the sentence by revising to include the word that is being modified. Consider the following examples with the modifier italicized:

> *Having walked five miles*, this bench will be the place to rest. (Incorrect; this version of the sentence implies that the bench walked the miles, not the person.)
>
> *Having walked five miles,* Matt will rest on this bench. (Correct; in this version, *having walked five miles* correctly modifies *Matt*, who did the walking.)

Since midnight, my dreams have been pleasant and comforting. (Incorrect; in this version, the adverb clause *since midnight* cannot modify the noun *dreams*.)

Since midnight, I have had pleasant and comforting dreams. (Correct; in this version, *since midnight* modifies the verb *have had*, telling us when the dreams occurred.)

Sometimes the modifier is not located close enough to the word it modifies for the sentence to be clearly understood. In this case, we call the error a misplaced modifier. Here is an example with the modifier italicized and the modified word in underlined.

We gave the hot <u>cocoa</u> to the children *that was filled with marshmallows.* (Incorrect; this sentence implies that the children are what are filled with marshmallows.)

We gave the hot <u>*cocoa*</u> *that was filled with marshmallows* to the children. (Correct; here, the cocoa is filled with marshmallows. The modifier is near the word it modifies.)

Parallelism and Subordination

Parallelism

To be grammatically correct we must use articles, prepositions, infinitives, and introductory words for dependent clauses consistently throughout a sentence. This is called parallelism. We use parallelism when we are matching parts of speech, phrases, or clauses with another part of the sentence. Being inconsistent creates confusion. Consider the following example.

Incorrect: Be ready for running and to ride a bike during the triathlon.

Correct: Be ready to run and to ride a bike during the triathlon.

Correct: Be ready for running and for riding a bike during the triathlon.

In the incorrect example, the gerund *running* does not match with the infinitive *to ride*. Either both should be infinitives or both should be gerunds.

Subordination

Sometimes we have unequal pieces of information in a sentence where one piece is more important than the other. We need to show that one piece of information is subordinate to the other. We can make the more important piece an independent clause and connect the other piece by making it a dependent clause. Consider this example:

Central thought: Kittens can always find their mother.

Subordinate: Kittens are blind at birth.

Complex Sentence: Despite being blind at birth, kittens can always find their mother.

The sentence "Kittens are blind at birth" is made subordinate to the sentence "Kittens can always find their mother" by placing the word "Despite" at the beginning and removing the subject, thus turning an independent clause ("kittens are blind at birth") into a subordinate phrase ("Despite being blind at birth").

Sentence Logic

Clauses

Clauses are groups of words within a sentence that have both a subject and a verb. We can distinguish a clause from a phrase because phrases do not have both a subject and a verb. There are several types of clauses; clauses can be independent or dependent and can serve as a noun, an adjective, or an adverb.

An *independent clause* could stand alone as its own sentence if the rest of the sentence were not there. For example:

> *The party is on Tuesday* after the volleyball game is over.
>
> *I am excited to go to the party* because my best friend will be there.

A *dependent clause*, or subordinating clause, is the part of the sentence that gives supportive information but cannot create a proper sentence by itself. However, it will still have both a subject and a verb; otherwise, it is a phrase. In the example above, *after the volleyball game is over* and *because my best friend will be there* are dependent because they begin with the conjunctions *after* and *because*, and a proper sentence does not begin with a conjunction.

Noun clauses are groups of words that collectively form a noun. Look for the opening words *whether, which, what, that, who, how,* or *why.* For example:

> I had fun cooking *what we had for dinner last night.*
>
> I'm going to track down *whoever ate my sandwich.*

Adjective clauses collectively form an adjective that modifies a noun or pronoun in the sentence. If you can remove the adjective clause and the leftovers create a standalone sentence, then the clause should be set off with commas, parentheses, or dashes. If you can remove the clause it is called nonrestrictive. If it can't be removed without ruining the sentence, then it is called restrictive and does not get set off with commas.

> Jenna*, who hates to get wet,* fell into the pool. (Nonrestrictive)
>
> The girl *who hates to get wet* fell into the pool. (Restrictive; the clause tells us which girl, and if removed there is confusion)

Adverbial clauses serve as an adverb in the sentence, modifying a verb, adjective, or other adverb. Look for the opening words *after, before, as, as if, although, because, if, since, so, so that, when, where, while,* or *unless.*

> She lost her wallet after she left the theme park.
>
> Her earring fell through the crack before she could catch it.

Phrases

A phrase is a group of words that go together but do not include both a subject and a verb. We use them to add information, explain something, or make the sentence easier for the reader to understand. Unlike clauses, phrases cannot ever stand alone as their own sentence if the rest of the sentence were not there. They do not form complete thoughts. There are noun phrases, prepositional phrases, verbal phrases, appositive phrases, and absolute phrases. Let's look at each of these.

Noun phrases: A noun phrase is a group of words built around a noun or pronoun that serves as a unit to form a noun in the sentence. Consider the following examples. The phrase is built around the underlined word. The entire phrase can be replaced by a noun or pronoun to test whether or not it is a noun phrase.

> I like the chocolate chip ice cream. (I like it.)
>
> I know all the shortest routes. (I know them.)
>
> I met the best supporting actress. (I met her.)

Prepositional phrases: These are phrases that begin with a preposition and end with a noun or pronoun. We use them as a unit to form the adjective or adverb in the sentence. Prepositional phrases that introduce a sentence are called introductory prepositional phrases and are set off with commas.

> I found the Frisbee *on the roof peak.* (Adverb; where it was found)
>
> The girl *with the bright red hair* was prom queen. (Adjective; which girl)
>
> *Before the sequel,* we wanted to watch the first movie. (Introductory phrase)

Verbal phrases: Some phrases look like verbs but do not serve as the verb in the sentence. These are called verbal phrases. There are three types: participial phrases, gerund phrases, and infinitive phrases.

Participial phrases start with a participle and modify nouns or pronouns; therefore, they act as the adjective in the sentence.

> *Beaten by the sun,* we searched for shade to sit in. (Modifies the pronoun *we*)
>
> The hikers, *being eaten by mosquitoes,* longed for repellant. (Modifies the noun *hikers*)

Gerund phrases often look like participles because they end in *-ing*, but they serve as the noun, not the adjective, in the sentence. Like any noun, we can use them as the subject or as the object of a verb or preposition in the sentence.

> *Eating green salad* is the best way to lose weight. (Subject)
>
> Sumo wrestlers are famous for *eating large quantities of food.* (Object)

Infinitive phrases often look like verbs because they start with the word *to,* but they serve as an adjective, adverb, or noun.

> *To survive the chill* is the goal of the Polar Bear Plunge. (Noun)
>
> A hot tub is at the scene *to warm up after the jump.* (Adverb)
>
> The jumpers have hot cocoa *to drink right away.* (Adjective)

Appositive phrases: We can use any of the above types of phrases to rename nouns or pronouns, and we call this an appositive phrase. Appositive phrases usually appear either just before or just after the noun or pronoun they are renaming. Appositive phrases are essential when the noun or pronoun is too general, and they are nonessential when they just add information.

> *The two famous brothers* Orville and Wilbur Wright invented the airplane. (Essential)
>
> Sarah Calysta, *my great grandmother,* is my namesake. (Nonessential)

Absolute phrases: When a participle comes after a noun and forms a phrase that is not otherwise part of the sentence, it's called an absolute phrase. Absolute phrases are not complete thoughts and cannot stand alone because they do not have a subject and a verb. They are not essential to the sentence in that they do not explain or add additional meaning to any other part of the sentence.

> *The engine roaring,* Jada closed her eyes and waited for the plane to take off.
>
> *The microphone crackling,* the flight attendant announced the delayed arrival.

Practice Questions

Read the selection and answer questions 1-5.

[1]I have to admit that when my father bought an RV, I thought he was making a huge mistake. [2]In fact, I even thought he might have gone a little bit crazy. [3]I did not really know anything about recreational vehicles, but I knew that my dad was as big a "city slicker" as there was. [4]On trips to the beach, he preferred to swim at the pool, and whenever he went hiking, he avoided touching any plants for fear that they might be poison ivy. [5]Why would this man, with an almost irrational fear of the outdoors, want a 40-foot camping behemoth?

[6]The RV was a great purchase for our family and brought us all closer together. [7]Every morning we would wake up, eat breakfast, and broke camp. [8]We laughed at our own comical attempts to back The Beast into spaces that seemed impossibly small. [9]We rejoiced when we figured out how to "hack" a solution to a nagging technological problem. [10]When things inevitably went wrong and we couldn't solve the problems on our own, we discovered the incredible helpfulness and friendliness of the RV community. We even made some new friends in the process.

[11] Above all, owning the RV allowed us to share adventures travelling across America that we could not have experienced in cars and hotels. [12]Enjoying a campfire on a chilly summer evening with the mountains of Glacier National Park in the background, or waking up early in the morning to see the sun rising over the distant spires of Arches National Park are memories that will always stay with me and our entire family. [13]Those are also memories that my siblings and I have now shared with our own children.

1. How should the author change sentence 11?
    a. Above all, it will allow us to share adventures travelling across America that we could not have experienced in cars and hotels.
    b. Above all, it allows you to share adventures travelling across America that you could not have experienced in cars and hotels.
    c. Above all, it allowed us to share adventures travelling across America that we could not have experienced in cars and hotels.
    d. Above all, it allows them to share adventures travelling across America that they could not have experienced in cars and hotels.

2. Which of the following examples would make a good addition to the selection after sentence 4?
    a. My father is also afraid of seeing insects.
    b. My father is surprisingly good at starting a campfire.
    c. My father negotiated contracts for a living.
    d. My father isn't even bothered by pigeons.

3. Which of the following would correct the error in sentence 7?
    a. Every morning we would wake up, ate breakfast, and broke camp.
    b. Every morning we would wake up, eat breakfast, and broke camp.
    c. Every morning we would wake up, eat breakfast, and break camp.
    d. Every morning we would wake up, ate breakfast, and break camp.

4. What transition word could be added to the beginning of sentence 6?
   a. Not surprisingly,
   b. Furthermore,
   c. As it turns out,
   d. Of course,

5. Which of the following topics would fit well between paragraph 1 and paragraph 2?
   a. A guide to RV holding tanks
   b. Describing how RV travel is actually not as outdoors-oriented as many think
   c. A description of different types of RVs
   d. Some examples of how other RV enthusiasts helped the narrator and his father during their travels

6. Which of the following is a clearer way to describe the following phrase?
   "employee-manager relations improvement guide"
   a. A guide to employing better managers
   b. A guide to improving relations between managers and employees
   c. A relationship between employees, managers, and improvement
   d. An improvement in employees' and managers' use of guides

Read the sentences, and then answer the following question.

7. Polls show that more and more people in the US distrust the government and view it as dysfunctional and corrupt. Every election, the same people are voted back into office.

Which word or words would best link these sentences?
   a. Not surprisingly,
   b. Understandably,
   c. And yet,
   d. Therefore,

8. Which of the following statements would make the best conclusion to an essay about civil rights activist Rosa Parks?
   a. On December 1, 1955, Rosa Parks refused to give up her bus seat to a white passenger, setting in motion the Montgomery bus boycott.
   b. Rosa Parks was a hero to many and came to symbolize the way that ordinary people could bring about real change in the Civil Rights Movement.
   c. Rosa Parks died in 2005 in Detroit, having moved from Montgomery shortly after the bus boycott.
   d. Rosa Parks' arrest was an early part of the Civil Rights Movement and helped lead to the passage of the Civil Rights Act of 1964.

Select the best version of the underlined part of the sentence. If you think the original sentence is best, choose the first answer.

9. Since <u>none of the furniture were delivered</u> on time, we have to move in at a later date.
   a. none of the furniture were delivered
   b. none of the furniture was delivered
   c. all of the furniture were delivered
   d. all of the furniture was delivered

10. An important issues stemming from this meeting is that we won't have enough time to meet all of the objectives.

a. An important issues stemming from this meeting
b. Important issue stemming from this meeting
c. An important issue stemming from this meeting
d. Important issues stemming from this meeting

11. There were many questions about what causes the case to have gone cold, but the detective wasn't willing to discuss it with reporters.

a. about what causes the case to have gone cold
b. about why the case is cold
c. about what causes the case to go cold
d. about why the case went cold

Directions for questions 12-16: Select the best version of the underlined part of the sentence. The first choice is the same as the original sentence. If you think the original sentence is best, choose the first answer.

12. The fact the train set only includes four cars and one small track was a big disappointment to my son.

a. the train set only includes four cars and one small track was a big disappointment
b. that the trains set only include four cars and one small track was a big disappointment
c. that the train set only includes four cars and one small track was a big disappointment
d. that the train set only includes four cars and one small track were a big disappointment

13. The rising popularity of the clean eating movement can be attributed to the fact that experts say added sugars and chemicals in our food are to blame for the obesity epidemic.

a. to the fact that experts say added sugars and chemicals in our food are to blame for the obesity epidemic.
b. in the facts that experts say added sugars and chemicals in our food are to blame for the obesity epidemic.
c. to the fact that experts saying added sugars and chemicals in our food are to blame for the obesity epidemic.
d. with the facts that experts say added sugars and chemicals in our food are to blame for the obesity epidemic.

14. She's looking for a suitcase that can fit all of her clothes, shoes, accessory, and makeup.

a. clothes, shoes, accessory, and makeup.
b. clothes, shoes, accessories, and makeup.
c. clothes, shoes, accessories, and makeups.
d. clothes, shoe, accessory, and makeup.

15. Because Shaun was used to playing guitar, he needs to work much harder at playing bass.

a. Because Shaun was used to playing guitar,
b. Even though Shaun is used to playing guitar,
c. While Shaun was used to playing guitar,
d. Because Shaun is used to playing guitar,

16. Considering the recent rains we have had, it's a wonder the plants haven't drowned.
    a. Considering the recent rains we have had, it's a wonder
    b. Consider the recent rains we have had, it's a wonder
    c. Considering for how much recent rain we have had, it's a wonder
    d. Considering, the recent rains we have had, it's a wonder

Directions for questions 17-20: Rewrite the sentence in your head following the directions given below. Keep in mind that your new sentence should be well written and should have essentially the same meaning as the original sentence.

17. There are many risks in firefighting, including smoke inhalation, exposure to hazardous materials, and oxygen deprivation, so firefighters are outfitted with many items that could save their lives, including a self-contained breathing apparatus.

Rewrite, beginning with so, firefighters.

The next words will be which of the following?
    a. are exposed to lots of dangerous situations.
    b. need to be very careful on the job.
    c. wear life-saving protective gear.
    d. have very risky jobs.

18. Though social media sites like Facebook, Instagram, and Snapchat have become increasingly popular, experts warn that teen users are exposing themselves to many dangers such as cyberbullying and predators.

Rewrite, beginning with experts warn that.

The next words will be which of the following?
    a. Facebook is dangerous.
    b. they are growing in popularity.
    c. teens are using them too much.
    d. they can be dangerous for teens.

19. Student loan debt is at an all-time high, which is why many politicians are using this issue to gain the attention and votes of students, or anyone with student loan debt.

Rewrite, beginning with Student loan debt is at an all-time high.

The next words will be which of the following?
    a. because politicians want students' votes.
    b. , so politicians are using the issue to gain votes.
    c. , so voters are choosing politicians who care about this issue.
    d. , and politicians want to do something about it.

20. Seasoned runners often advise new runners to get fitted for better quality running shoes because new runners often complain about minor injuries like sore knees or shin splints.

Rewrite, beginning with <u>Seasoned runners often advise new runners to get fitted for better quality running shoes.</u>

The next words will be which of the following?

a. to help them avoid minor injuries.
b. because they know better.
c. , so they can run further.
d. to complain about running injuries.

Answer Explanations

1. C: The sentence should be in the same tense and person as the rest of the selection. The rest of the selection is in past tense and first person. Choice *A* is in future tense. Choice *B* is in second person. Choice *D* is in third person. While none of these sentences are incorrect by themselves, they are written in a tense that is different from the rest of the selection. Only *Choice C* maintains tense and voice consistent with the rest of the selection.

2. A: Choices *B* and *D* go against the point the author is trying to make—that the father is not comfortable in nature. Choice *C* is irrelevant to the topic. Choice *A* is the only choice that emphasizes the father's discomfort with spending time in nature.

3. C: This sentence uses verbs in a parallel series, so each verb must follow the same pattern. In order to fit with the helping verb "would," each verb must be in the present tense. In Choices *A*, *B*, and *D*, one or more of the verbs switches to past tense. Only Choice *C* remains in the same tense, maintaining the pattern.

4. C: In paragraph 2, the author surprises the reader by asserting that the opposite of what was expected was in fact true—the city slicker father actually enjoyed the RV experience. Only Choice *C* indicates this shift in expected outcome, while the other choices indicate a continuation of the previous expectation.

5. B: Choices *A* and *C* are irrelevant to the topic. They deal more with details about RVs while the author is more concerned with the family's experiences with them. Choice *D* is relevant to the topic, but it would fit better between paragraphs 2 and 3, since the author does not mention this point until the end of the second paragraph. Choice *B* would help explain to the reader why the father, who does not enjoy the outdoors, could end up enjoying RVs so much.

6. B: Stacked modifying nouns such as this example are untangled by starting from the end and adding words as necessary to provide meaning. In this case, a *guide* to *improving relations* between *managers* and *employees*. Choices *C* and *D* do not define the item first as a guide. Choice *A* does identify as a guide, but confuses the order of the remaining descriptors. Choice *B* is correct, as it unstacks the nouns in the correct order and also makes logical sense.

7. C: The second sentence tells of an unexpected outcome of the first sentence. Choice *A*, Choice *B*, and Choice *D* indicate a logical progression, which does not match this surprise. Only Choice *C* indicates this unexpected twist.

8. B: Choice *A*, Choice *C*, and Choice *D* all relate facts but do not present the kind of general statement that would serve as an effective summary or conclusion. Choice *B* is correct.

9. B: Answer Choice *A* uses the plural form of the verb, when the subject is the pronoun *none*, which needs a singular verb. *C* also uses the wrong verb form and uses the word *all* in place of *none*, which doesn't make sense in the context of the sentence. *D* uses *all* again, and is missing the comma, which is necessary to set the dependent clause off from the independent clause.

10. C: In this answer, the article and subject agree, and the subject and predicate agree. Answer Choice *A* is incorrect because the article *an* and *issues* do not agree in number. *B* is incorrect because an article is needed before *important issue*. *D* is incorrect because the plural subject *issues* does not agree with the singular verb *is*.

11. D: Choices *A* and *C* use additional words and phrases that aren't necessary. *B* is more concise, but uses the present tense of *is*. This does not agree with the rest of the sentence, which uses past tense. The best choice is *D*, which uses the most concise sentence structure and is grammatically correct.

12. C: Choice *A* is missing the word *that*, which is necessary for the sentence to make sense. Choice *B* pluralizes *trains* and uses the singular form of the word *include*, so it does not agree with the word *set*. Choice *D* changes the verb to *were*, which is in plural form and does not agree with the singular subject.

13. A: Choices *B* and *D* both use the expression *attributed to the fact* incorrectly. It can only be attributed *to* the fact, not *with* or *in* the fact. Choice *C* incorrectly uses a gerund, *saying,* when it should use the present tense of the verb *say*.

14. B: Choice *B* is correct because it uses correct parallel structure of plural nouns. Choice *A* is incorrect because the word *accessory* is in singular form. Choice *C* is incorrect because it pluralizes *makeup*, which is already in plural form. Choice *D* is incorrect because it again uses the singular *accessory*, and it uses the singular *shoe*.

15. D: In a cause/effect relationship, it is correct to begin with the word *because*. This can eliminate both Choices *B* and *C*, which don't clearly show the cause/effect relationship. Choice *A* is incorrect because it uses the past tense, when the main clause is in the present tense. It makes grammatical sense for both parts of the sentence to be in present tense.

16. A: In answer Choice *B*, the present tense form of the verb *consider* creates an independent clause joined to another independent clause with only a comma, which is a comma splice and grammatically incorrect. Choices *C* and *D* use the possessive form of *its*, when it should be the contraction *it's* for *it is*. Choice *D* also includes incorrect comma placement.

17. C: The original sentence states that firefighting is dangerous, making it necessary for firefighters to wear protective gear. The portion of the sentence that needs to be rewritten focuses on the gear, not the dangers, of firefighting. *A*, *B*, and *D* all discuss the danger, not the gear, so *C* is the correct answer.

18. D: The original sentence states that though the sites are popular, they can be dangerous for teens, so *D* is the best choice. Choice *A* does state that there is danger, but it doesn't identify teens and limits it to just one site. Choice *B* repeats the statement from the beginning of the sentence, and *C* says the sites are used too much, which is not the point made in the original sentence.

19. B: The original sentence focuses on how politicians are using the student debt issue to their advantage, so *B* is the best answer choice. Choice *A* says politicians want students' votes, but suggests that it is the reason for student loan debt, which is incorrect. Choice *C* shifts the focus to voters, when the sentence is really about politicians. Choice *D* is vague and doesn't best restate the original meaning of the sentence.

20. A: This answer best matches the meaning of the original sentence, which states that seasoned runners offer advice to new runners because they have complaints of injuries. Choice *B* may be true, but it doesn't mention the complaints of injuries by new runners. Choice *C* may also be true, but it does not match the original meaning of the sentence. Choice *D* does not make sense in the context of the sentence.

Five-Paragraph Persuasive Essay

Brainstorming

One of the most important steps in writing an essay is prewriting. Before drafting an essay, it's helpful to think about the topic for a moment or two, in order to gain a more solid understanding of what the task is. Then, spending about five minutes jotting down the immediate ideas that could work for the essay is recommended. It is a way to get some words on the page and offer a reference for ideas when drafting. Scratch paper is provided for writers to use any prewriting techniques such as webbing, free writing, or listing. The goal is to get ideas out of the mind and onto the page.

Considering Opposing Viewpoints

In the planning stage, it's important to consider all aspects of the topic, including different viewpoints on the subject. There are more than two ways to look at a topic, and a strong argument considers those opposing viewpoints. Considering opposing viewpoints can help writers present a fair, balanced, and informed essay that shows consideration for all readers. This approach can also strengthen an argument by recognizing and potentially refuting the opposing viewpoint(s).

Drawing from personal experience may help to support ideas. For example, if the goal for writing is a personal narrative, then the story should be from the writer's own life. Many writers find it helpful to draw from personal experience, even in an essay that is not strictly narrative. Personal anecdotes or short stories can help to illustrate a point in other types of essays as well.

Moving from Brainstorming to Planning

Once the ideas are on the page, it's time to turn them into a solid plan for the essay. The best ideas from the brainstorming results can then be developed into a more formal outline. An outline typically has one main point (the thesis) and at least three sub-points that support the main point. Here's an example:

Main Idea

- Point #1
- Point #2
- Point #3

Of course, there will be details under each point, but this approach is the best for dealing with timed writing.

Staying on Track

Basing the essay on the outline aids in both organization and coherence. The goal is to ensure that there is enough time to develop each sub-point in the essay, roughly spending an equal amount of time on each idea. Keeping an eye on the time will help. If there are fifteen minutes left to draft the essay, then it makes sense to spend about 5 minutes on each of the ideas. Staying on task is critical to success, and timing out the parts of the essay can help writers avoid feeling overwhelmed.

Parts of the Essay

The introduction has to do a few important things:

- Establish the topic of the essay in original wording (i.e., not just repeating the prompt)
- Clarify the significance/importance of the topic or purpose for writing (not too many details, a brief overview)
- Offer a thesis statement that identifies the writer's own viewpoint on the topic (typically one-two brief sentences as a clear, concise explanation of the main point on the topic)

Body paragraphs reflect the ideas developed in the outline. Three-four points is probably sufficient for a short essay, and they should include the following:

- A topic sentence that identifies the sub-point (e.g., a reason why, a way how, a cause or effect)
- A detailed explanation of the point, explaining why the writer thinks this point is valid
- Illustrative examples, such as personal examples or real-world examples, that support and validate the point (i.e., "prove" the point)
- A concluding sentence that connects the examples, reasoning, and analysis to the point being made

The conclusion, or final paragraph, should be brief and should reiterate the focus, clarifying why the discussion is significant or important. It is important to avoid adding specific details or new ideas to this paragraph. The purpose of the conclusion is to sum up what has been said to bring the discussion to a close.

Don't Panic!

Writing an essay can be overwhelming, and performance panic is a natural response. The outline serves as a basis for the writing and helps writers keep focused. Getting stuck can also happen, and it's helpful to remember that brainstorming can be done at any time during the writing process. Following the steps of the writing process is the best defense against writer's block.

Timed essays can be particularly stressful, but assessors are trained to recognize the necessary planning and thinking for these timed efforts. Using the plan above and sticking to it helps with time management. Timing each part of the process helps writers stay on track. Sometimes writers try to cover too much in their essays. If time seems to be running out, this is an opportunity to determine whether all of the ideas in the outline are necessary. Three body paragraphs are sufficient, and more than that is probably too much to cover in a short essay.

More isn't always better in writing. A strong essay will be clear and concise. It will avoid unnecessary or repetitive details. It is better to have a concise, five-paragraph essay that makes a clear point, than a ten-paragraph essay that doesn't. The goal is to write one to two pages of quality writing. Paragraphs should also reflect balance; if the introduction goes to the bottom of the first page, the writing may be going off-track or be repetitive. It's best to fall into the one-two page range, but a complete, well-developed essay is the ultimate goal.

The Final Steps

Leaving a few minutes at the end to revise and proofread offers an opportunity for writers to polish things up. Putting one's self in the reader's shoes and focusing on what the essay actually says helps writers identify problems—it's a movement from the mindset of writer to the mindset of editor. The goal is to have a clean, clear copy of the essay. The following areas should be considered when proofreading:

- Sentence fragments
- Awkward sentence structure
- Run-on sentences
- Incorrect word choice
- Grammatical agreement errors
- Spelling errors
- Punctuation errors
- Capitalization errors

The Short Overview

The essay may seem challenging, but following these steps can help writers focus:

1. Take one or two minutes to think about the topic.
2. Generate some ideas through brainstorming (three-four minutes).
3. Organize ideas into a brief outline, selecting just three-four main points to cover in the essay (eventually the body paragraphs).
4. Develop essay in parts:
    a. Introduction paragraph, with intro to topic and main points
        i. Viewpoint on the subject at the end of the introduction
    b. Body paragraphs, based on outline
        i. Each paragraph: makes a main point, explains the viewpoint, uses examples to support the point
    c. Brief conclusion highlighting the main points and closing
5. Read over the essay (last five minutes).
    a. Look for any obvious errors, making sure that the writing makes sense.